# HRW

# ADVANCED ALGEBRA

## TEST GENERATOR: ASSESSMENT ITEM LISTING

Y0-CAD-588

explore

*communicate*

APPLY

$g(\theta)=15\sin(200\theta)$

$h(t)=-16t^2+27t+240$

**HOLT, RINEHART AND WINSTON**
*Harcourt Brace & Company*
**Austin** • New York • Orlando • Atlanta • San Francisco • Boston • Dallas • Toronto • London

ISBN 0-03-095296-4

3   4   5   6   7   066   99   98   97

# HRW
## TEST GENERATOR
## & ASSESSMENT
## ITEM LISTING

## SO CONVENIENT TOGETHER, YOU CAN...

&

### FLIP THE PAGE TO:

- Quickly preview the quantity, quality, and variety of **every** question, answer, and illustration in the *Test Generator* software package

- Assemble a list of questions you intend to use before accessing the *Test Generator* software

- Pick from a variety of questions, including quantitative comparison, multiple choice, and short answer

- Select questions by objective, chapter, or specific lesson

### CLICK THE MOUSE TO:

- Create worksheets, quizzes, mid-terms, and final exams

- Produce make-up exams or alternate tests using our handy editing capabilities

- Scramble question order

- Review your tests, including graphics

- Select questions by number, preview, random selection— or write them yourself

# TABLE OF CONTENTS

About the *HRW ADVANCED ALGEBRA Assessment Item Listing* and *Test Generator*     **v**

**Lesson Objectives**     **vi**

**Test Item Printout**

# About the *HRW Advanced Algebra Assessment Item Listing* and *Test Generator*

This *Assessment Item Listing* contains a complete printout of the test questions and answers from the *HRW Advanced Algebra Test Generator*. With the *Assessment Item Listing,* you can quickly preview the variety of test items stored in the *Test Generator* to assemble a list of questions you intend to use for worksheets, quizzes, and tests.

The *User's Guide* that is included in the *Test Generator* package contains directions for installing and using the program.

## Chapter Files

The test questions are arranged by chapter. Test questions are stored in 14 separate chapter files, one for each chapter in the textbook. Combinations of questions from any chapter may be used to generate a test.

## Question Attributes

Each test question is classified by attributes such as specific type, specific lesson, and lesson objective. The *Test Generator* lets you select questions based on these attributes. See the *User's Guide* for instructions.

## Question Types

Each chapter contains three different types of questions: Quantitative Comparison (**Q**), Multiple Choice (**M**), and Short Answer (**S**). When the program displays information about a question, it uses the one-letter abbreviation indicated.

## Lesson Number

Each test question is coded with the section number that links the question to its respective lesson in the chapter.

## Objective-Based Option

Each test question has an objective code that links the question to a lesson objective. Every lesson has one or two objectives, which are listed on the first page of the lesson in the *HRW Advanced Algebra Teacher's Edition*. See pages vi–xii for a complete listing of objectives.

## How to Get Started

To get started, install the program on your computer's hard disk. The *User's Guide* describes how to perform this procedure in the Installation section.

## Technical Support

The *Test Generator User's Guide* should answer any questions you may have about the program. However, if you need assistance, call the Holt, Rinehart and Winston Technical Support Center at 1-800-323-9239.

## Hardware Requirements

**Macintosh®:**
Macintosh® computer with a minimum of 800 KB of random access memory (RAM) available for the program, Macintosh® system software version 6.X or later, hard disk drive and diskette drive, printer

**IBM®:**
IBM® PC computer or any 100 percent compatible computer, Windows® Version 3.1 or later, hard drive and diskette drive, printer

# LESSON OBJECTIVES

## CHAPTER 1

**1a**    Identify linear equations and linear relationships of variables in a table.

**1b**    Write an equation describing a linear relationship, and graph it.

**2a**    Write an equation in slope-intercept form, given two points on the line or the slope and a point on the line.

**3a**    Graph a scatter plot from data in a table, and identify the correlation.

**3b**    Use a graphics calculator to find the correlation coefficient and to make predictions using the line of best fit.

**4a**    Solve problems involving direct variation.

**5a**    Solve problems by writing and solving linear equations.

**6a**    Solve and graph inequalities.

## CHAPTER 2

**1a**    Compare and identify number systems.

**1b**    Identify properties of real numbers, and use these properties to perform operations with rational numbers.

**2a**    Perform operations, and evaluate expressions using the properties of exponents.

**3a**    Identify and compare relations and functions, and use the vertical-line test to identify functions from their graphs.

**3b**    Use functions to model real-world applications, and give appropriate domain and range restrictions for the situation.

**4a**    Use function notation to define, evaluate, and operate with functions.

**5a**    Use the slope formula to write and identify increasing and decreasing linear functions.

**6a**    Identify and use properties of functions to add, subtract, multiply, and divide functions.

# CHAPTER 3

**1a** Identify the image and pre-image points and the axis of symmetry of a set of ordered pairs.

**1b** Determine the relationship of coordinates of points reflected over the $y$-axis, the $x$-axis, and the line $y = x$.

**2a** Determine the inverse of a function.

**2b** Define the inverse of a function, and use the horizontal-line test to determine whether the inverse is a function.

**3a** Define the composition of functions, and describe the relationship between the dependent and independent variables of functions that are composed.

**3b** Identify special properties of composition, and use these properties to analyze functions.

**3c** Develop a composition test to determine whether two functions are inverses.

**4a** Define and graph translations, reflections, and scalar transformations of the absolute value function.

**4b** Solve equations involving absolute value symbols by using graphing and algebraic methods.

**5a** Define and graph the greatest integer and the rounding-up functions.

**5b** Define step functions, and use them to model real-world applications.

**6a** Define and graph a system of parametric equations, and use them to model real-world applications.

**6b** Determine the linear function represented by a system of parametric equations.

# CHAPTER 4

**1a** Use matrices to store and represent data.

**1b** Add and subtract matrices.

**2a** Perform matrix multiplication.

**3a** Identify each type of system of two linear equations.

**3b** Solve a system of two linear equations using elimination by adding.

**4a** Represent a system of linear equations with an augmented matrix.

**4b** Solve a system of linear equations using the row reduction method and back substitution.

**5a** Find the inverse of a matrix.

**6a** Use matrix algebra to solve a system of equations.

**7a** Use matrices to represent and transform objects.

**8a** Graph the solution to a system of linear inequalities.

**9a** Graph a feasible region determined by constraints.

**10a** Find the maximum and minimum values of the objective function determined by the feasible region.

# CHAPTER 5

**1a** Write quadratic functions as a product of two linear functions that model real-world situations.

**1b** Use the graphics calculator to approximate the minimum or maximum value of a quadratic function and the $x$-intercepts of its graph.

**2a** Solve quadratic equations by taking square roots, or by squaring.

**2b** Use the distance formula to find the distance between two points.

**3a** Analyze graphs of quadratic functions to identify the transformations that result from changing the terms of the function.

**3b** Find the vertex, axis of symmetry, and direction of opening for the graphs of quadratic functions in the form $f(x) = a(x - h)^2 + k$.

**4a** Use tiles to complete the square for quadratic expressions of the form $x^2 + bx + c$, where $b$ and $c$ are real numbers.

**4b** Solve quadratic equations by completing the squares for expressions of the form $ax^2 + bx + c$, where $a \neq 0$, and $b$ and $c$ are real numbers.

**5a** Use the quadratic formula to solve quadratic equations that model real-world situations.

**5b** Use the axis of symmetry formula to find maximum or minimum values of quadratic equations that model real-world situations.

**6a** Determine the number of real-number solutions using the discriminant.

**6b** Solve quadratic equations with imaginary-number solutions.

**7a** Identify, operate with, and graph complex numbers.

**7b** Find the complex roots of quadratic equations that model real-world situations.

**8a** Write a quadratic model that fits three data points from real-world data.

**9a** Write, solve, and graph quadratic inequalities that model real-world situations.

# CHAPTER 6

**1a** Determine realistic domain restrictions of a volume function.

**1b** Define and examine the two forms of polynomial functions.

**2a** Write the factored form of a polynomial.

**2b** Define and use the Factor Theorem to find the zeros of polynomial functions.

**3a** Define and use the division algorithm for polynomials.

**3b** Define and use the Fundamental Theorem of Algebra to find the zeros of a polynomial function.

**4a** Determine and classify the behavior of single zeros and multiple zeros of a polynomial function.

**4b** Determine and classify the behavior and basic shape of polynomial functions of varying degrees.

**5a** Determine and analyze polynomial functions that model real-world situations.

**5b** Use the algebraic tool of variable substitution to simplify a polynomial function.

# CHAPTER 7

**1a**    Model the growth or decline of a population with a graphics calculator by repeatedly applying the multiplier.

**2a**    Identify the behavior of exponential functions by inspection and by graphing.

**2b**    Determine the growth of funds under various compounding methods.

**3a**    Identify the exponential and logarithmic functions as inverse functions.

**3b**    Determine equivalent forms for exponential and logarithmic equations.

**4a**    Identify the product, quotient, and power properties of logarithms.

**4b**    Simplify expressions and solve equations involving logarithms.

**5a**    Identify and use the common logarithmic function.

**5b**    Write equivalent logarithmic and exponential equations.

**6a**    Evaluate expressions involving the natural number, $e$, and identify the relationship between the natural logarithmic and exponential functions.

**6b**    Model growth and decay processes with natural exponential functions.

**7a**    Solve logarithmic and exponential equations by graphing.

**7b**    Use the exponential-log inverse properties to solve logarithmic and exponential equations.

# CHAPTER 8

**1a**    Identify the trigonometric ratios in special right triangles.

**1b**    Apply the special right triangle relationships to find missing lengths of sides of special triangles.

**2a**    Identify the coordinates of a point $(x, y)$ on the unit circle given an angle in standard position using the relationship $(x, y) = (\cos \theta, \sin \theta)$.

**2b**    Identify angles and their coterminal angles in standard position.

**3a**    Determine sides or angles of right triangles using trigonometric functions or their inverses.

**3b**    Determine, from the principal values of the inverse trigonometric functions, any angle in standard position.

**4a**    Identify how $a$, $b$, $c$, and $d$ in $f(\theta) = a + b \sin c(\theta - d)$ and $f(\theta) = a + b \cos c(\theta - d)$ transform the graphs of $f(\theta) = \sin \theta$ and $f(\theta) = \cos \theta$.

**5a**    Convert degree to radian measure, and radian to degree measure.

**5b**    Graph and identify properties of trigonometric functions in radian measure.

**6a**    Find the arc length and sector area determined by the central angle of a circle.

**7a**    Model applications with circular functions of the form $f(x) = a + b \cos c(x - d)$ or $f(x) = a + b \sin c(x - d)$.

# CHAPTER 9

**1a**    Given an inverse variation relationship, find the constant of variation, and write the equation of variation.

**2a**    Identify the graph, and write the equation of the vertical and horizontal asymptotes of a rational function that is the quotient of two linear functions.

**3a**    Determine whether a function is even, odd, or neither.

**3b**    Graph a function that is the reciprocal of a polynomial, and find its vertical asymptotes.

**4a**    Find the vertical and horizontal asymptotes, and the domain and range of a rational function in which the degree of the numerator is not greater than the degree of the denominator.

**5a**    Combine rational functions to get a single rational function.

**5b**    Solve equations that contain rational expressions.

# CHAPTER 10

**1a**    Write an equation of a parabola when given any two of the following: focus, directrix, vertex.

**1b**    Identify the vertex, focus, and directrix of a parabola from its equation; then, sketch a graph.

**2a**    Write the equation of a circle when given the coordinates of the center and the length of the radius.

**2b**    Determine the center and radius of a circle when given an equation of the circle.

**3a**    Determine the coordinates of the center, vertices, co-vertices, and foci when given an equation that represents an ellipse with either a horizontal or vertical major axis.

**4a**    Determine the coordinates of the center and foci, and find the lengths of the axes when given an equation that represents a hyperbola.

**5a**    Determine the intersections of a system consisting of one first-degree equation and one second-degree equation.

**5b**    Determine the intersections of a system consisting of two second-degree equations.

**6a**    Write the rectangular and parametric forms of the equation of a circle.

**6b**    Write the rectangular and parametric forms of the equation of an ellipse.

# CHAPTER 11

**1a**    Determine the theoretical probability of an event when given an appropriate sample space.

**1b**    Determine the experimental probability of an event when given the results of an experiment.

**2a**    Use the Fundamental Principal of Counting to determine how many ways a decision can be made.

**2b**    Determine the number of permutations of $n$ distinct objects taken $r$ at a time.

**3a**    Find the number of distinct permutations of $n$ objects of which $r_1$ objects are alike and $r_2$ objects are alike.

**3b**    Find the number of circular permutations of $n$ objects.

**4a**    Determine the number of combinations of $n$ objects taken $r$ at a time.

**4b**    Use the method for counting combinations with the Fundamental Principle of Counting to determine how many ways a decision can be made.

**5a**    Find the probability that two independent events $A$ and $B$ will both occur.

**5b**    Find the probability that either event $A$ or event $B$ will occur.

**6a**    Determine the probability of an event occurring that depends on the probability of another event occurring.

**7a**    Use random number simulations to approximate the probability of an event.

# CHAPTER 12

**1a**    Determine whether a given sequence is an arithmetic sequence.

**1b**    Determine whether a given sequence is a geometric sequence.

**2a**    Find the $n$th term of an arithmetic sequence.

**2b**    Find the sum of the first $n$ terms of an arithmetic series.

**3a**    Find the $n$th term of a geometric sequence.

**3b**    Find the sum of the first $n$ terms of an geometric series.

**4a**    Find the sum of an infinite geometric series.

**5a**    Use Pascal's Triangle to find combinations.

**6a**    Use the Binomial Theorem to expand $(a + b)^n$.

**6b**    Use the Binomial Theorem to find a particular term in the expansion of $(a + b)^n$.

# CHAPTER 13

**1a**   Find the measures of central tendency, mean, median, and mode of a given set of data.

**1b**   Construct a frequency table by dividing the data into classes, and find the class mean for the data.

**2a**   Construct a histogram for a given set of data.

**2b**   Construct a stem-and-leaf plot to represent a given set of data.

**3a**   Find the quartiles and interquartile range for a given set of data.

**3b**   Construct a box-and-whisker plot to display a given set of data.

**4a**   Find the range of a given set of data.

**4b**   Determine the mean deviation and the standard deviation for a given set of data.

**5a**   Find the probability of $r$ successes in $n$ trials of a binomial experiment.

**6a**   Given a set of data that is normally distributed, find the probability of an event if the mean and the standard deviation are known.

# CHAPTER 14

**1a**   Find all the parts of a triangle when given the measure of two angles and the length of the included side.

**2a**   Find all the parts of a triangle when given the lengths of two sides and the measure of the included angle.

**2b**   Find all the parts of a triangle when given the lengths of all three sides.

**3a**   Find the value of any of the other five trigonometric functions when given the value of one function and the quadrant in which the angle lies.

**4a**   Find exact values of the trigonometric functions by using the angle sum and difference identities, and the double angle identies.

**5a**   Solve equations containing trigonometric functions.

# CHAPTER 1

## QUANTITATIVE COMPARISON

In the space provided, write:
a. if the quantity of Column A is greater than the quantity in Column B;
b. if the quantity in Column B is greater than the quantity in Column A;
c. if the two quantities are equal; or
d. if the relationship cannot be determined from the information given.

| Column A | Column B | Answer |
|---|---|---|

1.

| slope of $y = x$ | $y$-intercept of $y = x$ | _____ |
|---|---|---|

2.

| the slope of the line passing through the points (-2, -3) and (2, 3) | the slope of the line passing through the points (-2, -2) and (3, 3) | _____ |
|---|---|---|

3.

| slope of $y = 27 + 4x$ | slope of $y = 27x + 4$ | _____ |
|---|---|---|

## MULTIPLE CHOICE   Circle the letter of the best answer choice.

4. In which of the following tables are the variables linearly related?

| a. | | b. | | c. | | d. | |
|---|---|---|---|---|---|---|---|
| $x$ | $y$ | $x$ | $y$ | $x$ | $y$ | $x$ | $y$ |
| 1 | 2 | -5 | 12 | -10 | 0 | 8 | 15 |
| 4 | 5 | -3 | 10 | -4 | 1 | 13 | 19 |
| 7 | 9 | -1 | 8 | 2 | -1 | 18 | 24 |
| 10 | 12 | 1 | 6 | 8 | 4 | 23 | 30 |
| 13 | 16 | 3 | 4 | 14 | -4 | 28 | 37 |
| 16 | 20 | 5 | 2 | 20 | 9 | 33 | 45 |

5. The values in the following table are linearly related.

| $x$ | $y$ |
|---|---|
| 410 | 512 |
| 1410 | $p$ |
| 2410 | 312 |
| 3410 | $q$ |

Which statement is true?
a. $p > q$     b. $q > p$         c. $p = q$         d. $p + q = 200$

6. A taxi company charges an initial fee of $2 per ride, plus $.50 for each mile. Which equation describes the relation between miles $m$ and cost $c$ (in dollars)?

a. $\frac{1}{2}(c + m) = 2$     b. $m = 2c + \frac{1}{2}$     c. $m = 2 + \frac{1}{2}c$     d. $c = 2 + \frac{1}{2}m$

7. Which graph describes the linear equation $y = \dfrac{-1}{3}x + 2$?

a.
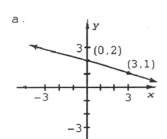

b.
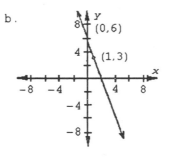

c.
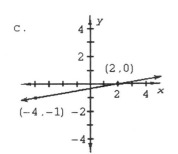

d.
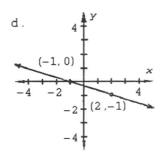

8. What is the equation of a line with slope $m = 0$ and $y$-intercept $\sqrt{5}$?

a. $y = \sqrt{5x}$    b. $x = \sqrt{5}$    c. $y = \sqrt{5}$      d. $y = x - \sqrt{5}$

9. The Castleville Company started in 1980 with 400 employees. Since then it has grown at a steady rate of 50 employees per year. Choose the linear equation that models the number of employees of the Castleville Company.

a. $y = 400 - 50x$      b. $y = 50 + 400x$

c. $y = 50x - 400$      d. $y = 400 + 50x$

10. Which of the following statements about this graph is true?
   a. The graph has no correlation.
   b. The graph has a positive correlation.
   c. The graph has a negative correlation.
   d. The line of best fit is $y = x$.

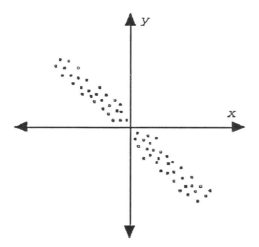

11. The following table shows the number of known deaths due to lightning in the continental United States during the years 1950-1960.

| Year | 1950 | 1951 | 1952 | 1953 | 1954 | 1955 | 1956 | 1957 |
|---|---|---|---|---|---|---|---|---|
| Known Lightning Deaths | 59 | 95 | 88 | 71 | 113 | 113 | 104 | 112 |

| Year | 1958 | 1959 | 1960 |
|---|---|---|---|
| Deaths | 73 | 153 | 100 |

Find the equation of the line of best fit using your graphics calculator. Using the line of best fit, how many lightning deaths do you predict in the continental United States in 1968?
a. 113          b. 73          c. 86          d. 153

12. Lengths of lines in a scale drawing vary directly with the actual lengths of lines in the subject of the drawing. In a scale drawing, a pool is 4 inches long and 3 inches wide. The pool is actually 24 feet wide. How long is the swimming pool?
a. 16 feet        b. 32 feet        c. 18 feet        d. 40 feet

13. Ohm's law, $V = IR$, states that the voltage, $V$, in an electric circuit varies directly with the electric current, $I$, measured in amperes. The constant of variation is the amount of resistance, $R$, in ohms. A toaster with 15 ohms of resistance is plugged into a 110-volt outlet. What current is carried by the toaster? Round your answer to the nearest tenth of an ampere.
a. 0.14 amperes     b. 7.3 amperes     c. 12.6 amperes     d. 95 ampere

14. The length of the hypotenuse of an isosceles right triangle varies directly as the length of one of the legs. Which equation describes the relationship between the lengths of the hypotenuse, $h$, and a leg, $l$?

a. $h = \dfrac{1}{\sqrt{2}}l$     b. $h = 2l$     c. $h = \dfrac{1}{2}l$     d. $h = \sqrt{2l}$

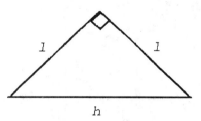

15. Solve by graphing:

$$1 - \frac{1}{2}x = 5 + \frac{1}{2}x$$

a. $x = -4$        b. $x = 4$        c. $x = \dfrac{-1}{3}$        d. $x = 3$

16. Given that $x + 3y = 14$, solve $25x - 7y = 22$ for $x$ and $y$.
a. $x = 4, y = 2$     b. $x = -5, y = 3$     c. $x = 8, y = 2$     d. $x = 2, y = 4$

17. Solve $rt = d$ for $t$.

a. $t = rd$     b. $t = \dfrac{d}{r}$     c. $t = \dfrac{r}{d}$     d. $t = d + r$

18. Solve $A = \dfrac{a + b + c + d}{4}$ for $c$.

a. $c = A = \dfrac{a + b + d}{4}$     b. $c = \dfrac{A - (a + b + d)}{4}$

c. $c = \dfrac{A - a + b + d}{4}$     d. $c = 4A - (a + b + d)$

3

19. Choose the inequality that describes the graph.

a. $y \geq \frac{3}{7}x + 3$    b. $y \leq \frac{3}{7}x + 3$    c. $y > \frac{3}{7}x + 3$    d. $y < \frac{3}{7}x + 3$

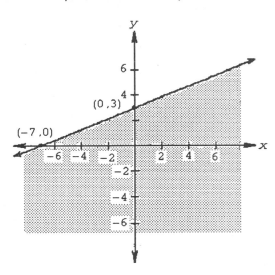

20. The perimeter of a triangle whose sides have lengths $x$, $2y$, and 6, cannot exceed 30 feet. Choose the linear inequality that describes this restriction.

a. $x + 2y \leq 6$    b. $x + 2y > 24$    c. $x + 2y \leq 24$    d. $x + 2y \geq 6$

21. Choose the inequality that describes the graph.

a. $x \leq 4$        b. $x \geq 4$            c. $x < 4$            d. $x > 4$

22. Choose the inequality that describes the graph.

a. $x \leq -3$        b. $x \geq -3$            c. $x < -3$            d. $x > -3$

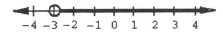

**SHORT ANSWER**   *Write the answer in the space provided.*

23. The variables in the following table are linearly related. What is the value of $a$?

| $x$ | .5 | 3 | 5.5 |
|---|---|---|---|
| $y$ | 15 | 6 | $a$ |

24. A department store tries selling a toy robot at different prices. The table shows the average number of robots sold per week at each price.

| price | average number sold per week |
|-------|------------------------------|
| $10   | 10                           |
| $12   | 9                            |
| $14   | 8                            |
| $15   | 7                            |
| $16   | 6                            |

Write a linear equation relating price $p$, in dollars, and average number $n$ of toy robots sold per week. At which of the five prices shown in the table is selling toy robots most profitable for the store?

_____

25. A music store charges $300 down and $50 per month to rent a cello. Write a linear equation relating months $m$ of cello rental and cost $c$ in dollars. Graph your equation.

_____

26. Each stair of a staircase is 8 inches tall and 1 foot wide. What is the slope of the line connecting the tops of the stairs in the diagram?

_____

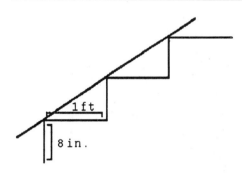

1 ft

8 in.

27. Write the equation of the graphed line in slope-intercept form.

_____

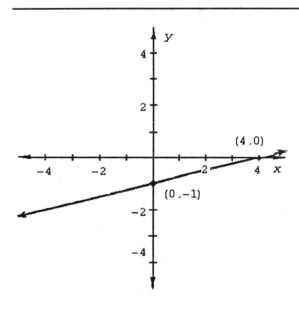

(4,0)

(0,-1)

28. Write the equation of a line which has the same slope as the line in the graph and a y-intercept of 1.

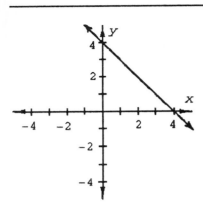

29. Write the equation of a line which has the same y-intercept as the line in the graph and a slope of 5.

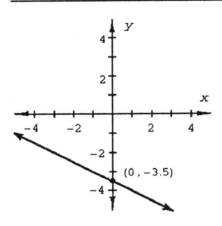

30. Graph the equation of the line that passes through the two points (2, 6) and (5, 3). Write the equation of the line in slope-intercept form.

31. Draw a scatter plot to represent the data in the following table.

| $x$ | 1 | 3 | 1 | 4 | 4 | 2 | 5 | 1 | 3 |
|---|---|---|---|---|---|---|---|---|---|
| $y$ | 10 | 9 | 11 | 7 | 8 | 9 | 6 | 3 | 8 |

Is the correlation between the variables $x$ and $y$ positive or negative?

32. The following table shows annual rainfall for several years in the same town.

| Year | 1985 | 1986 | 1987 | 1988 | 1989 | 1990 | 1991 | 1992 | 1993 |
|---|---|---|---|---|---|---|---|---|---|
| Rain (in.) | 35 | 31 | 28 | 47 | 32 | 36 | 34 | 48 | 25 |

Enter the data from the table into your graphics calculator. Find the correlation coefficient, $r$, to the nearest tenth.

33. The following table gives the height and weight of some adults.

| Height | 64" | 68" | 72" | 74" | 69" | 67" | 72" |
|--------|-----|-----|-----|-----|-----|-----|-----|
| Weight | 130 lb | 155 lb | 135 lb | 210 lb | 160 lb | 140 lb | 195 lb |

| Height | 62" | 70" | 63" | 73" | 71" | 68" | 60" |
|--------|-----|-----|-----|-----|-----|-----|-----|
| Weight | 110 lb | 180 lb | 170 lb | 190 lb | 160 lb | 160 lb | 105 lb |

Use your graphics calculator to find the correlation coefficient, $r$, and the equation of the line of best fit. Round all numerical coefficient to the nearest tenth.

_____

34. The following table gives sample ages and heights of some children.

| Age | 5 | 9 | 11 | 13 | 6 | 8 | 5 | 12 | 11 | 9 | 4 | 7 | 3 | 9 |
|-----|---|---|----|----|---|---|---|----|----|---|---|---|---|---|
| Height | 44" | 55" | 60" | 62" | 42" | 54" | 40" | 70" | 59" | 48" | 38" | 49" | 34" | 51" |

Find the line of best fit by using your graphing calculator. Use the line of best fit to predict the height of a 10-year-old child. Round your calculation to the nearest inch.

_____

35. The following table shows census figures for the population of the United States, rounded to the nearest million, for several different years. Use your graphics calculator to find the correlation coefficient and the equation of the line of best fit. Round all numbers to the nearest tenth.

| Year | 1900 | 1910 | 1920 | 1930 | 1940 | 1950 | 1960 | 1970 | 1980 | 1990 |
|------|------|------|------|------|------|------|------|------|------|------|
| Population | 76 | 92 | 106 | 123 | 132 | 151 | 179 | 203 | 227 | 249 |

_____

36. The following table shows the volumes and masses of the nine planets of earth's solar system (in units of earth volumes and earth masses). Use your graphics calculator to find the correlation coefficient, $r$, and the equation of the line of best fit. Round all numbers to the nearest hundredth.

| | Mercury | Venus | Earth | Mars | Jupiter | Saturn | Uranus | Neptune | Pluto |
|------|---------|-------|-------|------|---------|--------|--------|---------|-------|
| vol | 0.0559 | 0.8541 | 1.000 | 0.1506 | 1403 | 832 | 63 | 55 | 0.006 |
| mass | 0.0553 | 0.8150 | 1.000 | 0.1074 | 317.89 | 95.18 | 14.54 | 17.15 | 0.0020 |

_____

37. If $x$ varies directly as $y$, and $y$ is $\frac{3}{4}$ when $x$ is $\frac{4}{3}$, what is $y$ when $x$ is 16?

_____

38. If $x$ varies directly as $y$, and $y$ is 20.1 when $x$ is 0.67, what is $x$ when $y$ is –21?

_____

39. If $n$ varies directly as $m$, and $n$ is 52 when $m$ is –13. Find the constant of variation. Write the equation of direct variation.

_____

40. The law of gravitation states that weight varies directly with mass. A person with a mass of 70 kilograms weighs 154 pounds (on earth). Write the equation of variation that relates pounds $P$ to kilograms $K$. Graph the equation.

_____

41. Distances on a map vary directly with the actual distances in the region described by the map. On a map, Westwind is $4\frac{1}{2}$ inches from Shelby, and Shelby is 2 inches from Lyon. The distance from Westwind to Shelby is actually 135 miles. How many miles is Shelby from Lyon?

_____

42. Solve by graphing:

$3x + 4 = \frac{5}{2}$

_____

43. Solve by graphing:

$\frac{2}{3}x + 5 = \frac{1}{2}(x + 11)$

_____

44. Use your graphics calculator to estimate the solution to the nearest tenth.

$\frac{4}{7}x + 1 = \frac{7}{4}$

_____

45. Use your graphics calculator to estimate the solution to the nearest tenth.

$\frac{7}{8}x + \frac{14}{3} = 12$

_____

46. Given that $x = 4y - 21$, use substitution to solve $2x + 3y = 79$.

_____

47. Given that $y = \frac{-1}{2}x - \frac{1}{2}$, use substitution to solve $5y - 3x = 91$.

_____

48. The measure of one complementary angle is 1° less than six times the measure of its complement. Find the measure of each angle. Hint: the sum of an angle and its complement is 90°.

_____

49. Use your graphics calculator to graph the inequality $2x + 3y \leq 5$. Sketch the graph, labeling the $x$- and $y$-intercepts.

50. Sketch the graph of the inequality $\frac{3x}{2} + y < -1$.

51. Solve the inequality $\frac{1}{3}(5 - x) > 3$ and graph it on a number line.

_____

52. Solve the inequality $\dfrac{7-x}{8} \le \dfrac{5}{4}$ and graph it on a number line.

_____

53. Write an inequality that describes the graph.

_____

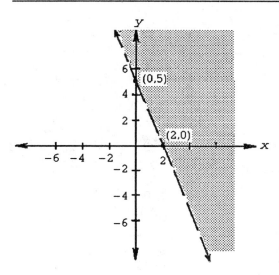

54. Solve and graph the inequality $2(y - x) \ge x + y + 1$

_____

55. Solve and graph the inequality.

$$-x + 2y \le -2\left[x + \dfrac{y}{2} + \dfrac{9}{2}\right]$$

_____

# *Answers to Chapter Questions*

1. Answer: a   Objective: 2A

2. Answer: a   Objective: 2a

3. Answer: b   Objective: 2a

4. Answer:
   b.

   | x | y |
   |----|----|
   | -5 | 12 |
   | -3 | 10 |
   | -1 | 8 |
   | 1 | 6 |
   | 3 | 4 |
   | 5 | 2 |

   Objective: 1a

5. Answer: a. $p > q$   Objective: 1a

6. Answer:
   d. $c = 2 + \dfrac{1}{2}$

   Objective: 1b

7. Answer:
   a.

   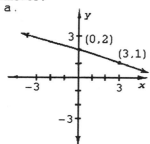

   Objective: 1b

8. Answer:
   c. $y = \sqrt{5}$

   Objective: 2a

9. Answer: d. $y = 400 + 50x$   Objective: 2a

10. Answer: c.   The graph has a negative correlation.   Objective: 3a

11. Answer: d. 153   Objective: 3b

12. Answer: b. 32 feet   Objective: 4a

13. Answer: b. 7.3 amperes   Objective: 4a

14. Answer:

    d. $h = \sqrt{21}$

    Objective: 4a

15. Answer: a. $x = 4$   Objective: 5a

16. Answer: d. $x = 2$, $y = 4$   Objective: 5a

17. Answer:

    b. $t = \dfrac{d}{r}$

    Objective: 5a

18. Answer: d. $c = 4A - (a + b + d)$   Objective: 5a

19. Answer:

    b. $y \le \dfrac{3}{7}x + 3$

    Objective: 6a

20. Answer: c. $x + 2y \le 24$   Objective: 6a

21. Answer: a. $x \le 4$   Objective: 6a

22. Answer: d. $x > -3$   Objective: 6a

23. Answer: $a = -3$   Objective: 1a

24. Answer:

    The linear equation is $n = \dfrac{-1}{2}p + b$. \$14 is the most profitable price shown.

    Objective: 1b

25. Answer: Equation: $c = 300 + 50m$

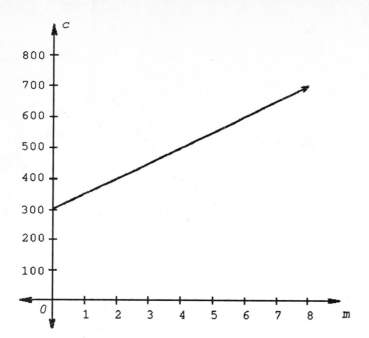

Objective: 1b

26. Answer:
$$\frac{2}{3}$$
Objective: 2a

27. Answer:
$$y = \frac{1}{4}x - 1$$
Objective: 2a

28. Answer: $y = -x + 1$.  Objective: 2a

29. Answer: $y = 5x - 3.5$  Objective: 2a

30. Answer: Equation: $y = -x + 8$

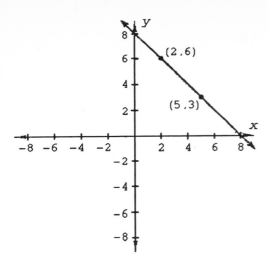

Objective: 2a

31. Answer: The correlation is negative.

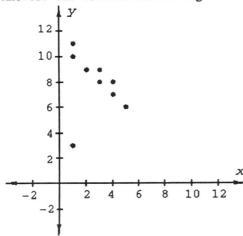

Objective: 3a

32. Answer: correlation coefficient: $r = .1$, to the nearest tenth.   Objective: 3b

33. Answer:
Correlation coefficient $r = .8$.
Equation of line of best fit: $y = 5.5x - 215.5$

Objective: 3b

34. Answer:
The line of best fit predicts about 57" as the height of a 10-year-old child. (equation of line: $y = 3.13x + 25.4$)

Objective: 3b

35. Answer:
   Correlation coefficient: $r = 1.0$
   Equation of best fit: $y = 1.9x - 3587.7$

   Objective: 3b


36. Answer:
   Correlation: $r = .96$
   Line of best fit: $y = .20x - 2.68$

   Objective: 3b


37. Answer: 9   Objective: 4a

38. Answer: −.7   Objective: 4a

39. Answer:
   Constant of variation: −4
   Equation: $n = -4m$

   Objective: 4a


40. Answer: $P = 2.2K$

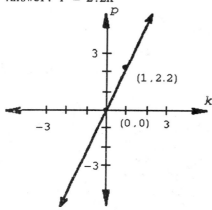

   Objective: 4a


41. Answer: 60 miles   Objective: 4a

42. Answer:
   $x = \dfrac{-1}{2}$

   Objective: 5a


43. Answer: $x = 3$   Objective: 5a

44. Answer: $x = 1.3$   Objective: 5a

45. Answer: $x = 8.4$   Objective: 5a

46. Answer: $x = 23$, $y = 11$   Objective: 5a

47. Answer: $x = -17$, $y = 8$   Objective: 5a

48. Answer: 77° and 13°   Objective: 5a

49. Answer:

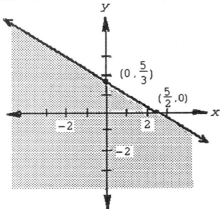

Objective: 6a

50. Answer:

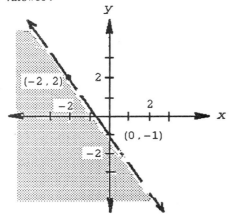

Objective: 6a

51. Answer: $x < -4$

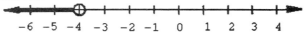

Objective: 6a

52. Answer: $x \geq -3$

Objective: 6a

53. Answer:

$y > \dfrac{-5}{2}x + 5$, or any equivalent inequality.

Objective: 6a

54. Answer: $y \geq 3x + 1$

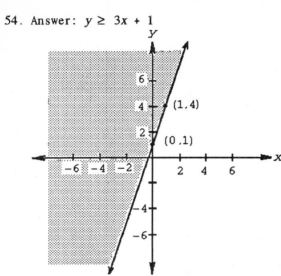

Objective: 6a

55. Answer:

$y \leq \dfrac{-1}{3}x - 3$

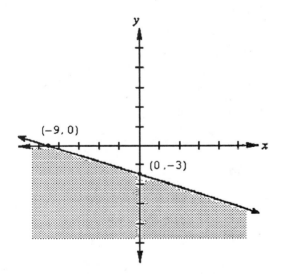

Objective: 6a

# CHAPTER 2

In the space provided, write:
a. if the quantity of Column A is greater than the quantity in Column B;
b. if the quantity in Column B is greater than the quantity in Column A;
c. if the two quantities are equal; or
d. if the relationship cannot be determined from the information given.

| Column A | Column B | Answer |
|---|---|---|

1.

| $(7^{11})\,(7^{-23})$ | $7^{-34}$ | _____ |
|---|---|---|

2.

| $\left[(-3)^4\right]^{-\frac{1}{4}}$ | $-\dfrac{1}{3}$ | _____ |
|---|---|---|

3.

| $9^2 - 8^2$ | $(9 - 8)^2$ | _____ |
|---|---|---|

**MULTIPLE CHOICE**   *Circle the letter of the best answer choice.*

4. _____ integers are rational numbers.
   a. All     b. Positive     c. Positive and negative     d. No

5. Simplify $\dfrac{x}{2y} - \dfrac{y}{2x}$.
   a. $\dfrac{1}{2}$     b. $\dfrac{x^2 - y^2}{4xy}$     c. $\dfrac{1}{4xy}$     d. $\dfrac{x^2 - y^2}{2xy}$

6. Simplify $\dfrac{12}{x} - \dfrac{13}{x^2}$.
   a. $\dfrac{12 - 13x}{156x^2}$     b. $\dfrac{156}{x^3}$     c. $\dfrac{-1}{x - x^2}$     d. $\dfrac{12x - 13}{x^2}$

7. Simplify $\dfrac{2(x - 7)}{x} \cdot \dfrac{5x^2}{4(x - 7)}$.
   a. $5x(x - 7)^2$     b. $\dfrac{5x}{2}$     c. $\dfrac{-5x}{2}$     d. $\dfrac{5x^2}{7 + x}$

8. Simplify $\dfrac{2x - 6}{x + 3} \cdot \dfrac{x + 15}{3x - 9}$.
   a. $\dfrac{2x + 30}{3x + 9}$     b. $\dfrac{2}{3}(x + 15)$     c. $\dfrac{10}{3}$     d. $\dfrac{-10}{3}$

9. Simplify $\left[\dfrac{-2x^3 y^{-4}}{x^7 y^{-2}}\right]^4$.
   a. $\dfrac{-16}{x^4 y^2}$     b. $\dfrac{-16x^{10}}{y^2}$     c. $\dfrac{16y^8}{x^{16}}$     d. $\dfrac{16}{x^{16} y^8}$

10.

Simplify $\left[\dfrac{3x^{-1}y^0}{4x^5y^{-2}}\right]^{-4}$.

a. $\dfrac{-12x^{24}}{y^8}$    b. $\dfrac{256x^{24}}{81y^8}$    c. $\dfrac{-256x^{10}y^6}{81}$    c. $\dfrac{81y^8}{256x^{24}}$

11.

Simplify $\left[x^{\frac{1}{3}}\right]\left[x^3\right]\left[x^{-2}\right]$.

a. $x^{\frac{2}{3}}$    b. $x^{\frac{-4}{3}}$    c. $x^{\frac{-2}{3}}$    d. $x^{\frac{4}{3}}$

12.

Simplify $\left[ab(a+b)^{-2}\right]^{-3}$.

a. $a^3 + b^3$    b. $(a^2b + b^2a)^6$    c. $(ab)^{-3}(a+b)^{-5}$    d. $(ab)^{-3}(a+b)^6$

13. Which are graphs of functions?

    a. I only     b. II only     c. I and II     d. II and III

I. 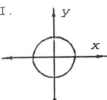     II. 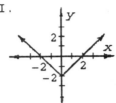     III.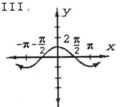

14. Which of the following tables represents a function?

| I. | | | II. | | | III. | |
|---|---|---|---|---|---|---|---|
| $x$ | $y$ | | $x$ | $y$ | | $x$ | $y$ |
| 0 | 1 | | 1 | 0 | | 1 | 1 |
| 1 | 1 | | 1 | 1 | | 1 | -1 |
| -1 | 1 | | 1 | -1 | | 4 | 2 |
| 2 | -1 | | -1 | 2 | | 4 | -2 |

   a. I only     b. II only     c. III only     d. I and III

15. Which sets of ordered pairs represent functions?

   I. {(2,1), (-2,1), (-7,5), (7,5)}
   II. {(3,-2), (-2,3), (3,-1), (-2,2)}
   III. {(-9,0), (11,0), (0,0), ($\frac{-2}{3}$,0)}

   a. I only     b. II only     c. III only     d. I and III

16. Let $h(x) = x^2$. What is $h(a+b)$?

   a. $a^2 + b^2$    b. $a^2 + 2ab + b^2$    c. $a2 - b^2$    d. $a^2 - 2ab + b^2$

17. Let $g(x) = 5x + 4$. Find $g(y^2 - y)$.

   a. $5y^2 - y + 8$    b. $5y^2 - y + 4$    c. $5y^2 - 5y + 4$    d. $5y^2 - 5y + 8$

18. Let $f(x) = x^2 - x$. What is $f(a+3)$?

   a. $a^2 - a - 3$    b. $a^2 - a$    c. $a^2 + 6x + 9$    d. $a^2 + 5a + 6$

19. Let $g(x) = 7x - 2$. Find $\dfrac{g(a) - g(b)}{a - b}$.

   a. 7     b. $7(a - b)$     c. $\dfrac{7a - 7b - 4}{a - b}$     d. $\dfrac{7a - 7b + 2}{a - b}$

20. Which graph represents an increasing linear function?

a.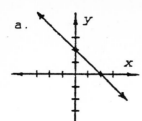

b.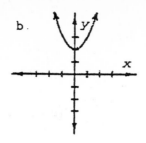

c.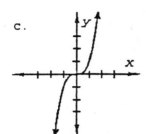

d.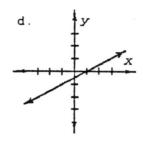

21. Let $f(x) = \dfrac{x - 2}{3}$, $g(x) = 14 - 7x$, $h(x) = \pi$, $k(x) = \pi x$.

Which functions are increasing functions of $x$?
a. $f$ and $g$     b. $g$ and $h$     c. $f$ and $h$      d. $f$ and $k$

22. Let $f(x) = 5x$ and $g(x) = \dfrac{-1}{2}$. Which graph represents the function $f \cdot g$?

a.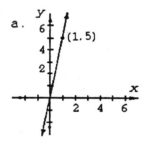

b.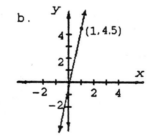

c.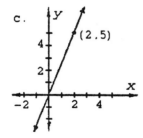

d.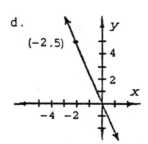

23. Let $f(x) = x^2 + 5x$. What is $f(x - 2)$?
    a. $x^2 + x - 6$     b. $x^2 + x + 4$     c. $x^2 + 5x - 2$     d. $x^2 + 5x - 10$

24. Let $f(x) = 2x^2 - 15x$. What is $-f(x)$?
    a. $15x - 2x^2$     b. $2x^2 + 15x$     c. $-2x^2 - 15x$     d. $2x^2 - 15x$

25. Simplify $\frac{5}{x} + \frac{x}{5}$.

_____

26. Simplify $\frac{y - 2}{7} - \frac{y + 3}{21}$

_____

27. Simplify $\frac{5a}{4b} \cdot \frac{12b^2}{35a^2}$.

_____

28. Simplify $\frac{x + 2}{x - 2} - 1$.

_____

29. Simplify $\frac{-3b^{-2}}{81b^{-10}}$.

_____

30. Simplify $(5x^{-3}y^{10})(4x^2y^4)^{\frac{-3}{2}}$.

_____

31. Simplify $\frac{x^{a-b}y^b}{x^{b-a}y^a}$.

_____

32. Roy has invested $400 at 5.5% interest, compounded once a year. The formula for the amount, $A$, of money he will have in $t$ years is $A = 400(1.055)^t$. What will be the value of the investment after one year? After 2 years?

_____

33. Using your graphics calculator find the domain and range of $y = 2|x| - x$.

_____

34. What is the range of the function $y = -\frac{x}{2}$ when the domain is $\{-1, 0, 1, 2\}$?

_____

35. A bicycle rental company charges $5 per day, plus a base fee of $20. Write a function relating cost and number of days rented.

_____

22

36. Use this function to determine the cost of renting a bicycle for 1 week (7 days).

_____

37. Using your graphics calculator find the domain and range of $y = -3x^2$.

_____

38. Using your graphics calculator find the domain and range of $y = \dfrac{1}{x^2}$.

_____

39. Let $f(x) = 4x^3 + 3x^2$. Find $f(-5)$.

_____

40. Use your graphics calculator to find the domain and range of $f(x) = x^2 - 6x + 13$.

_____

41. The base of a triangle is 5 more than $\dfrac{1}{3}$ of its height. Express the area of the triangle as a function of its height using function notation.

_____

42. Let $f(x) = 3|x| - x$. What is $f(-11)$?

_____

43. Let $g(x) = -x^2 + 2x + 2$. Use your graphics calculator to find the vertex of $f$.

_____

44. Let $h(x) = x^4 - x^3 + x$. What is $h(-2)$?

_____

45. Find the slope of the linear function containing the points $(-11, 4)$ and $(-10, 8)$, and write the function.

_____

46. Write a linear function $f$ such that $f(3) = 2$ and $f(-3) = 4$. Is this function increasing, decreasing, or neither?

_____

47. Write a linear function $f$ such that $f(0) = \sqrt{3}$ and the slope of $f$ is 0. Is f increasing, decreasing, or neither?

_____

48. Write the linear function represented by the graph. Is this function increasing, decreasing, or neither?

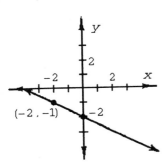

49. Write a linear function $f(x)$ such that $f(0) = \frac{-2}{3}$ and $f(x)$ increases by 5 when $x$ increases by 4.

50. Use your graphics calculator to graph $f(x) = 9 - x^2$. Find the $x$-intercepts of $f$. Is $f$ a linear function of $x$?

51. The population of the city of Tarnot was 6258 in 1990. In 1995, the population was 8758. The population of Tarnot is an increasing linear function of time. Write the function.

52. Use your graphics calculator to graph the function $f(x) = 2.4 - 3x$. Find the $x$-intercepts of $f$. . Is $f$ a linear function of $x$?

53. Let $f(x) = x^2 - x$ and $g(x) = x + 4$. What is $(f + g)(5)$?

54. Let $g(x) = x^2 + 3x - 2$ and $h(x) = -x^2 + 8$. Write the function $(g + h)(x)$.

55. Let $f(x) = |x|$ and let $g(x) = x$. Use your graphics calculator to graph the function $f \cdot \frac{1}{g}$. Find the domain and range of this function.

56. Let $f(x) = 3x - 2$. Write the function $-f(x)$. Graph $f(x)$ and $-f(x)$ on the same coordinate plane.

57. Let $f(x) = 3x^2 - 2x + 1$. What is $f(-x)$?

# *Answers to Chapter Questions*

1. Answer: a   Objective: 2A

2. Answer: c   Objective: 2a

3. Answer: a   Objective: 2a

4. Answer: a. All   Objective: 1a

5. Answer:
   d. $\dfrac{x^2 - y^2}{2xy}$

   Objective: 1b

6. Answer:
   d. $\dfrac{12x - 13}{x^2}$

   Objective: 1b

7. Answer:
   b. $\dfrac{5x}{2}$

   Objective: 1b

8. Answer:
   a. $\dfrac{2x + 30}{3x + 9}$

   Objective: 1b

9. Answer:
   d. $\dfrac{16}{x^{16}y^8}$

   Objective: 2a

10. Answer:
    b. $\dfrac{256x^{24}}{81y^8}$

    Objective: 2a

11. Answer:
    d. $x^{\frac{4}{3}}$

    Objective: 2a

12. Answer: d. $(ab)^{-3}(a + b)^6$   Objective: 2a

13. Answer: d. II and III   Objective: 3a

14. Answer: a. I only   Objective: 3a

15. Answer: d. I and III   Objective: 3a

16. Answer: b. $a^2 + 2ab + b^2$   Objective: 4a

17. Answer: c. $5y^2 - 5y + 4$   Objective: 4a

18. Answer: d. $a^2 + 5a + 6$   Objective: 4a

19. Answer: a. 7   Objective: 4a

20. Answer:

d.

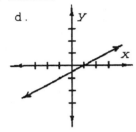

Objective: 5a

21. Answer: d. $f$ and $k$   Objective: 5a

22. Answer:

d.

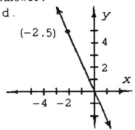

Objective: 6a

23. Answer: a. $x^2 + x - 6$   Objective: 6a

24. Answer: a. $15x - 2x^2$   Objective: 6a

25. Answer:
$$\frac{25 + x^2}{5x}$$

Objective: 1b

26. Answer:
$$\frac{2y - 9}{21}$$
Objective: 1b


27. Answer:
$$\frac{3b}{7a}$$
Objective: 1b


28. Answer:
$$\frac{4}{x - 2}$$
Objective: 1b


29. Answer:
$$\frac{-b^8}{27}$$
Objective: 2a


30. Answer:
$$\frac{5y^4}{8x^6}$$
Objective: 2a


31. Answer: $x^{2a - 2b}y^{b-a}$   Objective: 2a

32. Answer: After one year it will be \$422.00;   after two years it will be \$445.21.

Objective: 2a

33. Answer:
Domain: all real numbers
Range: all real numbers greater than or equal to 0
Objective: 3a


34. Answer:
$$\{\frac{1}{2}, 0, \frac{-1}{2}, -1\}$$
Objective: 3b


35. Answer: $y = 20 + 5x$   Objective: 3b

36. Answer: \$55   Objective: 3b

37. Answer:
Domain: all real numbers
Range: all real numbers $y \leq 0$
Objective: 3b

38. Answer:
Domain: all real numbers except 0
Range: all positive real numbers

Objective: 3b


39. Answer: -425  Objective: 4a

40. Answer:
Domain: all real numbers
Range: all real numbers greater than or equal to 4.

Objective: 4a


41. Answer:
$$f(h) = \frac{1}{2}h(5 + \frac{1}{3}h)$$

Objective: 4a


42. Answer: 44  Objective: 4a

43. Answer: (1, 3)  Objective: 4a

44. Answer: 22  Objective: 4a

45. Answer:
slope: 4
$f(x) = 4x + 48$

Objective: 5a


46. Answer:
$$f(x) = \frac{-1}{3}x + 3$$
This is a decreasing function of $x$.

Objective: 5a


47. Answer:
$$f(x) = \sqrt{3}$$
This function is neither increasing nor decreasing.

Objective: 5a


48. Answer:
$$f(x) = \frac{-1}{2}x - 2$$
This is a decreasing function of $x$.

Objective: 5a

49. Answer:
$$f(x) = \frac{5}{4}x - \frac{2}{3}$$
Objective: 5a

50. Answer:
$x$-intercepts: (3, 0) (-3, 0)
No, $f$ is not a linear function of $x$.

Objective: 5a

51. Answer: $f(x) = 500x - 998,742$   Objective: 5a

52. Answer:
The $x$-intercept of $f$ is (.8, 0).
Yes, $f$ is a linear function of $x$.

Objective: 5a

53. Answer: 29   Objective: 6a

54. Answer: $(g + h)x = 3x + 6$   Objective: 6a

55. Answer:
Domain: all real numbers except zero
Range: {-1.1}

Objective: 6a

56. Answer: $-f(x) = -3x + 2$

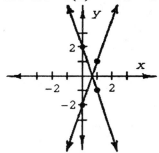

Objective: 6a

57. Answer: $f(-x) = 3x^2 + 2x + 1$   Objective: 6a

# CHAPTER 3

## QUANTITATIVE COMPARISON

In the space provided, write:
a. if the quantity of Column A is greater than the quantity in Column B;
b. if the quantity in Column B is greater than the quantity in Column A;
c. if the two quantities are equal; or
d. if the relationship cannot be determined from the information given.

| Column A | Column B | Answer |
|---|---|---|

1. Let $f(x) = 3x + 2$.

| $f(11)$ | $f^{-1}(11)$ | _____ |
|---|---|---|

2.

| $(f \circ g)(x)$ if $f$ is the inverse of $g$ | $(g \circ f)(x)$ if $f$ is the inverse of $g$ | _____ |
|---|---|---|

3.

| $5 - |-3| - |-8|$ | $5 + |-3| - |-8|$ | _____ |
|---|---|---|

4.

| $[-18.1]$ | $[-17.9]$ | _____ |
|---|---|---|

**MULTIPLE CHOICE**   *Circle the letter of the best answer choice.*

5. What image point is symmetric to the point (5, 3) with respect to the line $y = 2$?
   a. (-1, 3)    b. (5, 1)    c. (5, -3)    d. (3, 5)

6. Let $S$ be the set of ordered pairs {(-2, 0), (-1, 1), (0, -2), (1, 3)}. What set of image points is symmetric to $S$ with respect to the y-axis?
   a. {(-2, 0), (-1, -1), (0, 2), (1, -3)}
   b. {(1, -1), (0, 2), (-1, -3), (2, 0)}
   c. {(2, 0), (-1, 3), (1, 1), (0, -2)}
   d. {(0, -2), (2, 0), (1, -1), (3, 1)}

7. What image point is symmetric to the point (5, -7) with respect to the x-axis?
   a. (-5, -7)    b. (-5, 7)    c. (-7, 5)    d. (5, 7)

8. What image point is symmetric to the point $\left[\pi, \sqrt{2}\right]$ with respect to the y-axis?
   a. $\left[-\pi, \sqrt{2}\right]$    b. $\left[\pi, -\sqrt{2}\right]$    c. $\left[-\pi, -\sqrt{2}\right]$    d. $\left[\sqrt{2}, \pi\right]$

9. What image point is symmetric to the point $\left[\frac{1}{2}, -\frac{3}{7}\right]$ with respect to the line $y = x$?
   a. $\left[-\frac{1}{2}, -\frac{3}{7}\right]$    b. $\left[-\frac{3}{7}, -\frac{1}{2}\right]$    c. $\left[-\frac{3}{7}, \frac{1}{2}\right]$    d. $\left[\frac{1}{2}, \frac{3}{7}\right]$

10. Let $f(x) = \dfrac{2x - 5}{3}$. What is the inverse of $f$?

   a. $f^{-1}(x) = \dfrac{3}{2x - 5}$          b. $f^{-1}(x) = \dfrac{3x + 5}{2}$

   c. $f^{-1}(x) = 3\left[\dfrac{x}{2} + 5\right]$          d. $f^{-1}(x) = \dfrac{x}{3} + 5$

11. Let $f(x) = -\dfrac{5}{x}$. What is the inverse of $f$?

   a. $f^{-1}(x) = 5x$          b. $f^{-1}(x) = -\dfrac{x}{5}$

   c. $f^{-1}(x) = -\dfrac{5}{x}$          d. $f^{-1}(x) = -\dfrac{1}{5x}$

12. Which of the functions has an inverse function?
   a. I only      b. II only      c. III only      d. I and II

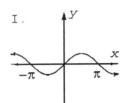 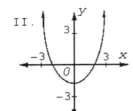 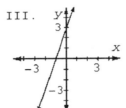

13. Use your graphics calculator and the horizontal line test to determine which of the following functions are invertible.
      I. $f(x) = x^2 - x$            II. $g(x) = -x^3$
      III. $h(x) = \pi$             IV. $k(x) = \pi x$
   a. I only      b. I and II      c. II only      d. II and IV

14. Let $h(x) = (x - 2)^2 - 4$. Choose two functions $f$ and $g$ such that $(g \circ f)(x) = h(x)$.
   a. $f(x) = x^2$; $g(x) = x - 6$        b. $f(x) = x - 4$; $g(x) = (x - 2)^2$
   c. $f(x) = x - 4$; $g(x) = x$          d. $f(x) = x - 2$; $g(x) = x^2 - 4$

15. What function does the graph represent?
   a. $f(x) = |x - 2|$            b. $f(x) = |3x - 2|$
   c. $f(x) = -\dfrac{1}{2}|x| - 2$        d. $f(x) = -\dfrac{1}{3}|x - 2|$

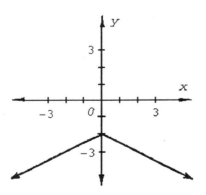

16. Let $f(x) = 3|x - 2|$. Find $f(-6)$.
   a. 12      b. -24      c. 24      d. -18

17. What function does the graph represent?
    a. $f(x) = [x]$
    b. $f(x) = [x + 1]$
    c. $f(x) = [x + 2]$
    d. $f(x) = [x - 1]$

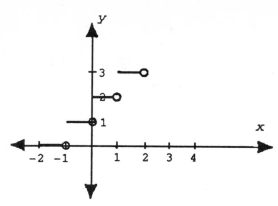

18. Which graph represents the function $f(x) = [x - 3]$?

a.

b.

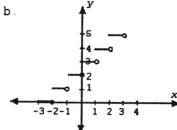

c.

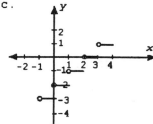

d.

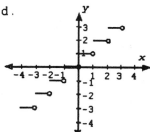

19. Simplify $\lceil \sqrt{3} \rceil$.
    a. 3
    b. 2
    c. -2
    d. 1

20. Combine the parametric system into one linear function.
    $\begin{cases} x(t) = 5 + t \\ y(t) = 5 - t \end{cases}$
    a. $f(x) = 10 - x$
    b. $f(x) = x - 10$
    c. $f(x) = x + 10$
    d. $f(x) = 2x$

21. Combine the parametric system into one linear equation.
    $\begin{cases} x(t) = t - \dfrac{5}{2} \\ y(t) = 2t + 4 \end{cases}$
    a. $f(x) = 5x - 4$
    b. $f(x) = 2x - 1$
    c. $f(x) = 2x + 9$
    d. $f(x) = \dfrac{2}{5}x - 5$

22. What system of parametric equations describes the line through (3, -3) and (6, -2)?

a. $\begin{cases} x = 5t \\ y = t - 6 \end{cases}$   b. $\begin{cases} x = \dfrac{3}{2}t \\ y = t + \dfrac{1}{2} \end{cases}$   c. $\begin{cases} x = 9t \\ y = 3t - 4 \end{cases}$   d. $\begin{cases} x = 3t \\ y = t - 4 \end{cases}$

23. Which graph describes the parametric system?
$$\begin{cases} x = 4t + 2 \\ y = 2t + 2 \end{cases}$$

a.
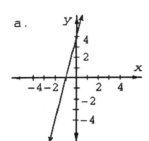

b.
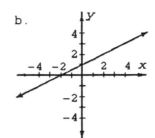

c.
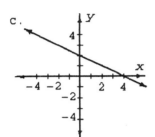

d.
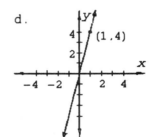

24. Combine the parametric system into one linear function.
$$\begin{cases} x(t) = 5t - 1 \\ y(t) = 2t + 3 \end{cases}$$

a. $y = \dfrac{2}{5}x + \dfrac{1}{3}$      b. $y = \dfrac{5}{2}x - 3$

c. $y = -4x - 14$      d. $y = \dfrac{2}{5}x + \dfrac{17}{5}$

**SHORT ANSWER** *Write the answer in the space provided.*

25. Plot the image points that are symmetric to the vertices of $\triangle ABC$ with respect to the $y$-axis. Then draw the triangle whose vertices are the points of the image set.

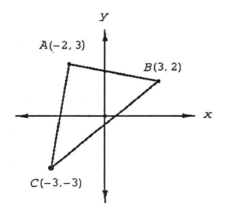

26. Plot the image points that are symmetric to the vertices of △PQR with respect to the x-axis. Then draw the triangle whose vertices are the points of the image set.

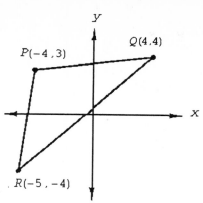

27. Plot the image points that are symmetric to the vertices of △WXY with respect to the line y = x. Then draw the triangle whose vertices are the points of the image set.

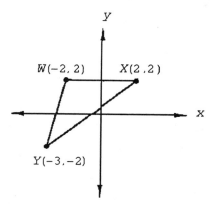

28. Plot the image points that are symmetric to the vertices of rectangle ABCD with respect to the line y = x. Then draw the rectangle whose vertices are the points of the image set.

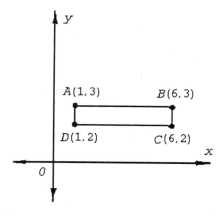

29. Plot the set of ordered pairs: {(2, 2), (1, 1), (4, 2), (5, 1)}. Then plot the image points that are symmetric to the set with respect to the y = x.

30. Let $f(x) = -\frac{2}{3}x + 7$. Find the inverse of $f$.

31. Let $f(x) = 15 - x$. Find the inverse of $f$.

_____

32. Find the inverse of $g(x) = 4 - \dfrac{1}{x}$.

_____

33. Let $h(x) = \dfrac{4x + 2}{7}$. What is $h^{-1}\left[\dfrac{22}{7}\right]$?

_____

34. Let $f(x) = 5x$ and let $g(x) = x + 3$. Find $f \circ g$ and $g \circ f$.

_____

35. Let $g(x) = \dfrac{1}{x}$ and let $h(x) = x - 2$. Find $g \circ h$ and $h \circ g$.

_____

36. Let $k(x) = \sqrt{x}$ and $h(x) = x - 64$. Find $(k \circ h)(100)$ and $(h \circ k)(100)$.

_____

37. Let $p(x) = |x|$ and let $q(x) = 10 - x$. Find $(p \circ q)(-15)$ and $(q \circ p)(-15)$.

_____

38. Let $f(x) = x - 1$ and $g(x) = x^2 + 7$. Find $g \circ f$. Use your graphics calculator to find the domain and range of $g \circ f$.

_____

39. The Axlerod Appliance Company sells refrigerators on the installment plan for a $100 down payment and $18 per month. Write a function that relates the amount paid, $A$, to the number of months, $x$. How many months would it take to pay for a refrigerator that costs $316?

_____

40. Let $f(x) = 2x - 8$ and $g(x) = \dfrac{1}{2}x - 4$. Determine whether the composition of the functions $f$ and $g$ is commutative.

_____

41. Graph $f(x) = -|x + 2|$. What transformations were applied to $f(x) = |x|$?

_____

42. Graph $f(x) = |x - 3|$. What transformations were applied to $f(x) = |x|$?

_____

43. Graph $f(x) = -\frac{1}{2}|x|$. What transformations were applied to $f(x) = |x|$?

_____

44. Graph $f(x) = |x| - 1$. What transformations were applied to $f(x) = |x|$?

_____

45. Simplify $|-7 + 2| - |7|$.

_____

46. Solve for $x$: $|12 - 3x| = 3$.

_____

47. Solve for $x$: $|4x - 3| = 2$.

_____

48. Graph the function $f(x) = [x] + 4$.

49. Let $x = 1.9$. Find $[3x]$, $3[x]$, $[-3x]$, and $-3[x]$.

_____

50. Simplify $[\sqrt{2}\,]$.

_____

51. Simplify $\left[-\frac{11}{3}\right]$.

_____

52. It costs $10 to rent a rowboat for up to an hour. For each additional hour, or fraction thereof, there is an added charge of $2.50. Write a step function for the cost, $C(t)$, of renting a rowboat for $t$ hours.

_____

53. A hotel charges $60 for a room for up to 1 day (24 hours), and $60 for each additional day, or fraction thereof. Write a step function for the cost $C(x)$ of renting a hotel room for $x$ days, where $x$ is any real number greater than zero.

_____

54. A train starts at a point 5 miles east and 12 miles north of the main station. Every hour it travels 35 miles east and 42 miles north. Write a parametric system for the train's position, with the main station as the origin, $x$ as the east-west coordinate, and $y$ as the north-south coordinate.

_____

55. Combine the parametric system into one linear function.
$$\begin{cases} x(t) = -\frac{t}{2} + 3 \\ y(t) = \frac{t}{4} \end{cases}$$

_____

56. Combine the parametric system into one linear function.
$$\begin{cases} x = 2t + 1 \\ y = 6t + 1 \end{cases}$$

_____

# *Answers to Chapter Questions*

1. Answer: a   Objective: 2B

2. Answer: c   Objective: 3c

3. Answer: b   Objective: 4b

4. Answer: b   Objective: 5b

5. Answer: b. (5, 1)   Objective: 1a

6. Answer: c. {(2, 0), (-1, 3), (1, 1), (0, -2)}   Objective: 1b

7. Answer: d. (5, 7)   Objective: 1b

8. Answer:
   a. $\left[-\pi, \sqrt{2}\right]$

   Objective: 1b


9. Answer:
   c. $\left[-\dfrac{3}{7}, \dfrac{1}{2}\right]$

   Objective: 1b


10. Answer:
    b. $f^{-1}(x) = \dfrac{3x + 5}{2}$

    Objective: 2a


11. Answer:
    c. $f^{-1}(x) = -\dfrac{5}{x}$

    Objective: 2a


12. Answer: c. III only   Objective: 2b

13. Answer: d. II and IV   Objective: 2b

14. Answer: d. $f(x) = x - 2$; $g(x) = x^2 - 4$   Objective: 3a

15. Answer:
    c. $f(x) = -\dfrac{1}{2}|x| - 2$

    Objective: 4a

16. Answer: c. 24  Objective: 4b

17. Answer: c. $f(x) = [x + 2]$  Objective: 5a

18. Answer:
    c.

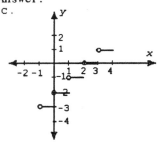

    Objective: 5a

19. Answer: b. 2  Objective: 5b

20. Answer: a. $f(x) = 10 - x$  Objective: 6b

21. Answer: c. $f(x) = 2x + 9$  Objective: 6b

22. Answer:
    d. $\begin{cases} x = 3t \\ y = t - 4 \end{cases}$

    Objective: 6b

23. Answer:
    b.

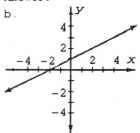

    Objective: 6b

24. Answer:
    d. $y = \frac{2}{5}x + \frac{17}{5}$

    Objective: 6b

25. Answer:

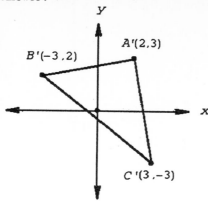

Objective: 1a

26. Answer:

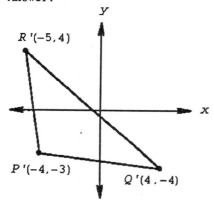

Objective: 1a

27. Answer:

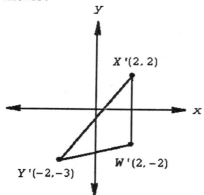

Objective: 1a

41

28. Answer:

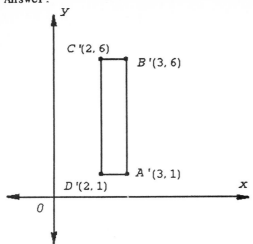

Objective: 1a

29. Answer:

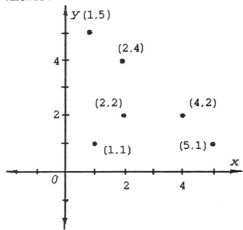

Objective: 1b

30. Answer:

$$f^{-1}(x) = \frac{21 - 3x}{2}$$

Objective: 2a

31. Answer: $f^{-1}(x) = 15 - x$   Objective: 2a

32. Answer:

$$g^{-1}(x) = \frac{1}{4 - x}$$

Objective: 2a

42

33. Answer: 5  Objective: 2a

34. Answer: $(f \circ g)(x) = 5x + 15$; $(g \circ f)(x) = 5x + 3$  Objective: 3a

35. Answer:

$(g \circ h)(x) = \dfrac{1}{x - 2}$; $(h \circ g)x = \dfrac{1}{x} - 2$.

Objective: 3a

36. Answer: $(k \circ h)100 = \pm 6$; $(h \circ k)(100) = -54$ or $-74$.  Objective: 3a

37. Answer: $(p \circ q)(-15) = 25$; $(q \circ p)(-15) = -5$.  Objective: 3a

38. Answer:
Domain: all real numbers. Range: all real numbers greater than or equal to 7.
$(g \circ f)(x) = (x - 1)^2 + 7$

Objective: 3b

39. Answer: $A(x) = 18x + 100$. It would take 12 months.  Objective: 3b

40. Answer: no  Objective: 3c

41. Answer: reflection in $x$-axis, horizontal translation of $-2$

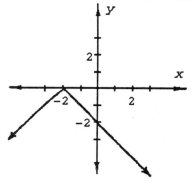

Objective: 4a

42. Answer: horizontal translation of 3

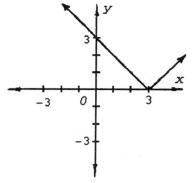

Objective: 4a

43. Answer:

reflection in *x*-axis, scalar transformation of $\frac{1}{2}$ (widen by a factor of 2)

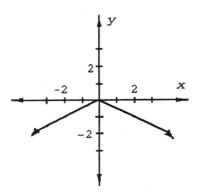

Objective: 4a

44. Answer: vertical translation of -1

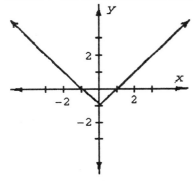

Objective: 4a

45. Answer: -2   Objective: 4b

46. Answer: $x = 5$ or $x = 3$.   Objective: 4b

47. Answer:

$x = \frac{5}{4}$ or $x = \frac{1}{4}$.

Objective: 4b

48. Answer:

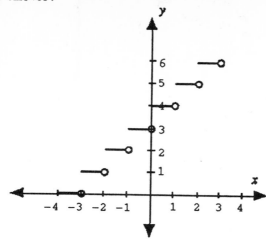

Objective: 5a

49. Answer: $[3x] = 5$; $3[x] = 3$; $[-3x] = -6$; $-3[x] = -3$  Objective: 5a

50. Answer: 1  Objective: 5b

51. Answer: -3  Objective: 5b

52. Answer:

$C(t) = 10 + 2.50\lceil t \rceil$, $t > 0$.

Objective: 5b

53. Answer:

$C(x) = 60\lceil x \rceil$

Objective: 5b

54. Answer:

$\begin{cases} x = 5 + 35t \\ y = 12 + 42t \end{cases}$

Objective: 6a

55. Answer:

$y = -\dfrac{x}{2} + \dfrac{3}{2}$

Objective: 6a

56. Answer: $f(x) = 3x - 2$  Objective: 6b

# CHAPTER 4

*QUANTITATIVE COMPARISON*

In the space provided, write:
a. if the quantity of Column A is greater than the quantity in Column B;
b. if the quantity in Column B is greater than the quantity in Column A;
c. if the two quantities are equal; or
d. if the relationship cannot be determined from the information given.

| Column A | Column B | Answer |
|---|---|---|

1.
$$J = \begin{bmatrix} 4 & 2 & 7 \\ 6 & 3 & 0 \\ 8 & 2 & 1 \end{bmatrix}$$

| $j_{13}$ | $j_{31}$ | _____ |
|---|---|---|

2.
$$J = \begin{bmatrix} 4 & 2 & 7 \\ 6 & 3 & 0 \\ 8 & 2 & 1 \end{bmatrix}$$

| $j_{12}$ | $j_{32}$ | _____ |
|---|---|---|

3.

| value of $V$ at $(2, 6)$ if $V = 4x + 12y$ | value of $V$ at $(6, 2)$ if $V = 4x + 12y$ | _____ |
|---|---|---|

*MULTIPLE CHOICE   Circle the letter of the best answer choice.*

4. What is $h_{3\,2}$ in matrix $H$?

$$H = \begin{bmatrix} 1 & -1 & 2 & -2 \\ 0 & \dfrac{1}{2} & -\dfrac{1}{6} & \dfrac{2}{3} \\ \dfrac{4}{3} & 3 & 0 & -1 \\ \dfrac{8}{9} & 1 & 5 & 7 \end{bmatrix}$$

a. $-\dfrac{1}{6}$　　　b. 3　　　c. 0　　　d. $\dfrac{1}{2}$

5. Let $V = \begin{bmatrix} 11 & -1 \\ 1 & 0 \end{bmatrix}$ and $W = \begin{bmatrix} -5 & 27 \\ -2 & -8 \end{bmatrix}$. What is $V + W$?

a. $\begin{bmatrix} 16 & -28 \\ 3 & 8 \end{bmatrix}$　　b. $\begin{bmatrix} -6 & 38 \\ -2 & -7 \end{bmatrix}$　　c. $\begin{bmatrix} -4 & 27 \\ 9 & -9 \end{bmatrix}$　　d. $\begin{bmatrix} 6 & 26 \\ -1 & -8 \end{bmatrix}$

6.

Let $A = \begin{bmatrix} 1 & 5 \\ 2 & 0.1 \end{bmatrix}$, $B = \begin{bmatrix} \pi & 0 \\ \sqrt{2} & -2 \\ -7 & \frac{22}{3} \end{bmatrix}$, and $C = \begin{bmatrix} \frac{4}{7} & 0 & 0 \\ 0 & \frac{9}{2} & 0 \\ 0 & 0 & -5 \end{bmatrix}$. Which products

can be found?

a. $AB$ and $BC$

b. $BA$ and $AC$

c. $BC$ and $CB$

d. $BA$ and $CB$

7.

Let $P = \begin{bmatrix} 1 & 3 \\ 0 & -9 \\ 2 & 0 \end{bmatrix}$ and $Q = \begin{bmatrix} 6 & 2 & 0 & 4 \\ -3 & 0 & 5 & -1 \end{bmatrix}$. Use your graphics calculator

to find the product $PQ$.

a. $\begin{bmatrix} -3 & 2 & 15 & 1 \\ 27 & 0 & -45 & 9 \\ 12 & 4 & 0 & 8 \end{bmatrix}$

b. $\begin{bmatrix} 12 & 2 & -15 & 1 \\ -27 & 0 & 45 & -9 \\ 0 & -4 & 0 & 8 \end{bmatrix}$

c. $\begin{bmatrix} 1 & 8 & 0 \\ -3 & 27 & 45 \\ 12 & 32 & 16 \\ -2 & 0 & 17 \end{bmatrix}$

d. none of these

8. A theater concession stand sells soft drinks in three sizes. Matrix $B$ shows the price of each size of soft drink. Matrix $A$ shows the number of each size of drink sold on Monday and Tuesday. How much money did the concession stand make from soft drink sales on Tuesday

$$\begin{array}{c} \text{Small} \quad \text{Medium} \quad \text{Large} \\ \begin{array}{c} \text{Monday} \\ \text{Tuesday} \end{array} \begin{bmatrix} 19 & 47 & 158 \\ 36 & 112 & 97 \end{bmatrix} = A \end{array} \qquad \begin{array}{c} \text{Price (\$)} \\ \begin{array}{c} \text{Small} \\ \text{Medium} \\ \text{Large} \end{array} \begin{bmatrix} 0.40 \\ 0.65 \\ 0.95 \end{bmatrix} = B \end{array}$$

a. \$188.25    b. \$145.80    c. \$111.80    d. \$179.35

9.

Let $A = \begin{bmatrix} 1 & 0 \\ 0 & 1 \end{bmatrix}$, $B = \begin{bmatrix} 4 & -2 \\ 6 & 0 \end{bmatrix}$, and $C = \begin{bmatrix} x & 2 \\ y & -13 \end{bmatrix}$. Which products can be

found?

a. $AB$ and $AC$    b. $BA$ and $BC$    c. $CB$ and $CA$    d. all of these

10. A baker sells chocolate chip cookies for \$1.40 per lb and double fudge cookies for \$2.40 per lb. She wants to make 30 lbs of a mixture of chocolate chip and double fudge cookies to sell for \$2.00 per lb. How many lbs of each kind of cookie should she use?

a. 15 lbs of each kind

b. 10 lbs of chocolate chip cookies and 20 lbs of double fudge cookies

c. 18 lbs of chocolate chip cookies and 12 lbs of double fudge cookies

d. 12 lbs of chocolate chip cookies and 18 lbs of double fudge cookies

11. Choose the augmented matrix for the system of equations.

$$\begin{cases} 5x + 12 = z - y \\ x + 3y = 9z \\ 2x + 7 = z \end{cases}$$

a. $\begin{bmatrix} 5 & 12 & 1 & : & -1 \\ 1 & 3 & 9 & : & 0 \\ 2 & 0 & 7 & : & 1 \end{bmatrix}$

b. $\begin{bmatrix} 5 & -1 & 1 & : & 12 \\ 1 & 3 & 9 & : & 0 \\ 2 & 0 & 1 & : & 7 \end{bmatrix}$

c. $\begin{bmatrix} 5 & 1 & -1 & : & -12 \\ 1 & 3 & -9 & : & 0 \\ 2 & 0 & -1 & : & -7 \end{bmatrix}$

d. $\begin{bmatrix} 5 & -1 & 1 & : & 12 \\ 1 & 3 & 9 & : & -9 \\ 2 & 0 & -1 & : & 7 \end{bmatrix}$

12. Let $J = \begin{bmatrix} -1 & 0 \\ 2 & 3 \end{bmatrix}$. Find $J^{-1}$, if it exists.

a. $\begin{bmatrix} -1 & 0 \\ \dfrac{2}{3} & \dfrac{1}{3} \end{bmatrix}$

b. $\begin{bmatrix} 1 & 0 \\ -\dfrac{2}{3} & \dfrac{1}{3} \end{bmatrix}$

c. $\begin{bmatrix} 3 & 2 \\ 0 & -1 \end{bmatrix}$

d. $J$ has no inverse.

13. Let $W = \begin{bmatrix} 4 & 2 \\ 1 & 1 \end{bmatrix}$. Find $W^{-1}$, if it exists.

a. $\begin{bmatrix} -2 & 1 \\ 1 & -4 \end{bmatrix}$

b. $W$ has no inverse.

c. $\begin{bmatrix} -1 & 2 \\ \dfrac{1}{2} & -\dfrac{1}{2} \end{bmatrix}$

d. $\begin{bmatrix} \dfrac{1}{2} & -1 \\ -\dfrac{1}{2} & 2 \end{bmatrix}$

14. Let $V = \begin{bmatrix} 5 & -2 \\ 10 & -4 \end{bmatrix}$. Find $V^{-1}$, if it exists.

a. $\begin{bmatrix} -\dfrac{1}{5} & \dfrac{1}{10} \\ -\dfrac{1}{2} & \dfrac{1}{4} \end{bmatrix}$

b. $\begin{bmatrix} \dfrac{1}{5} & -\dfrac{1}{4} \\ \dfrac{1}{2} & -\dfrac{1}{10} \end{bmatrix}$

c. $\begin{bmatrix} -4 & 2 \\ -10 & 5 \end{bmatrix}$

d. The matrix $V$ has no inverse.

15. Which of the following matrices are their own inverses?

I. $\begin{bmatrix} 2 & 0 \\ 0 & 2 \end{bmatrix}$    II. $\begin{bmatrix} 0 & 1 \\ -1 & 0 \end{bmatrix}$    III. $\begin{bmatrix} 0 & -1 \\ -1 & 0 \end{bmatrix}$

a. I only    b. I and II    c. II and III    d. III only

16. Choose the matrix equation that represents the system of equations.

$$\begin{cases} x - 3y = 27 \\ 11.2x + z = 4 \\ 9x - 5y + 7 = 56 \end{cases}$$

a. $\begin{bmatrix} 1 & -3 & 0 \\ 11.2 & 1 & 0 \\ 9 & -5 & 7 \end{bmatrix} \begin{bmatrix} x \\ y \\ z \end{bmatrix} = \begin{bmatrix} 27 \\ 4 \\ 56 \end{bmatrix}$     b. $\begin{bmatrix} 1 & -3 & 0 \\ 11.2 & 0 & 1 \\ 9 & -5 & 7 \end{bmatrix} \begin{bmatrix} x \\ y \\ x \end{bmatrix} = \begin{bmatrix} 27 \\ 4 \\ 56 \end{bmatrix}$

c. $\begin{bmatrix} 1 & -3 & 0 \\ 11.2 & 0 & 1 \\ 9 & -5 & 0 \end{bmatrix} \begin{bmatrix} x \\ y \\ z \end{bmatrix} = \begin{bmatrix} 27 \\ 4 \\ 63 \end{bmatrix}$     d. $\begin{bmatrix} 1 & -3 & 0 \\ 11.2 & 0 & 1 \\ 9 & -5 & 0 \end{bmatrix} \begin{bmatrix} x \\ y \\ z \end{bmatrix} = \begin{bmatrix} 27 \\ 4 \\ 49 \end{bmatrix}$

17. Choose the correct solution to the matrix equation.

$$\begin{bmatrix} 3 & 4 & 0 \\ 0 & 5 & 1 \\ 4 & 0 & -8 \end{bmatrix} \begin{bmatrix} x \\ y \\ z \end{bmatrix} = \begin{bmatrix} 110 \\ 94 \\ -16 \end{bmatrix}$$

a. $\begin{bmatrix} x \\ y \\ z \end{bmatrix} = \begin{bmatrix} 11 \\ 28 \\ 51 \end{bmatrix}$     b. $\begin{bmatrix} x \\ y \\ z \end{bmatrix} = \begin{bmatrix} 7 \\ 40 \\ 22 \end{bmatrix}$

c. $\begin{bmatrix} x \\ y \\ z \end{bmatrix} = \begin{bmatrix} 30 \\ 5 \\ 27 \end{bmatrix}$     d. $\begin{bmatrix} x \\ y \\ z \end{bmatrix} = \begin{bmatrix} 14 \\ 17 \\ 9 \end{bmatrix}$

18. On a hot day, a public swimming pool sold 248 tickets. Tickets cost $0.25 for children, $2.00 for adults under 65, and $1.00 for senior citizens. Three times as many children's tickets as senior citizen's tickets were sold. The total amount of money collected was $221.00. How many tickets for adults under 65 were sold?

a. 33          b. 144          c. 72          d. 36

19.

Let $M = \begin{bmatrix} -2 & 3 & 1 \\ 1 & 5 & -3 \end{bmatrix}$. What is the shape of the object represented by matrix $M$?

a. line segment     b. triangle     c. trapezoid     d. rectangle

20.

Let $C = \begin{bmatrix} 0 & 0 & 2 \\ 3 & 0 & 0 \end{bmatrix}$. Choose the matrix equation that describes counter-clockwise rotation by 90° of the object represented by $C$.

a. $\begin{bmatrix} 0 & 1 \\ -1 & 0 \end{bmatrix} \begin{bmatrix} 0 & 0 & 2 \\ 3 & 0 & 0 \end{bmatrix} = \begin{bmatrix} -3 & 0 & 0 \\ 0 & 0 & 2 \end{bmatrix}$     b. $\begin{bmatrix} 0 & 1 \\ 1 & 0 \end{bmatrix} \begin{bmatrix} 0 & 0 & 2 \\ 3 & 0 & 0 \end{bmatrix} = \begin{bmatrix} -3 & 0 & 0 \\ 0 & 0 & 2 \end{bmatrix}$

c. $\begin{bmatrix} 1 & 1 \\ -1 & 0 \end{bmatrix} \begin{bmatrix} 0 & 0 & 2 \\ 3 & 0 & 0 \end{bmatrix} = \begin{bmatrix} -3 & 0 & 0 \\ 0 & 0 & 2 \end{bmatrix}$     d. $\begin{bmatrix} 0 & -1 \\ 1 & 0 \end{bmatrix} \begin{bmatrix} 0 & 0 & 2 \\ 3 & 0 & 0 \end{bmatrix} = \begin{bmatrix} -3 & 0 & 0 \\ 0 & 0 & 2 \end{bmatrix}$

21.

Let $K = \begin{bmatrix} -1 & 1 & 0 \\ 2 & 0 & -1 \end{bmatrix}$. Choose the matrix that can be used to transform the object represented by $K$ by rotating it 90° clockwise and shrinking it to half its size

a. $\begin{bmatrix} 0 & -\frac{1}{2} \\ \frac{1}{2} & 0 \end{bmatrix}$
b. $\begin{bmatrix} 0 & -1 \\ 1 & 0 \end{bmatrix}$
c. $\begin{bmatrix} 0 & 1 \\ 1 & 0 \end{bmatrix}$
d. $\begin{bmatrix} 0 & \frac{1}{2} \\ -\frac{1}{2} & 0 \end{bmatrix}$

22.

Let $P = \begin{bmatrix} 3 & 5 & 1 \\ 5 & 3 & 1 \end{bmatrix}$. Choose the matrix that can transform $P$ by rotating it 180° and tripling its size.

a. $\begin{bmatrix} -\frac{1}{3} & 0 \\ 0 & \frac{1}{3} \end{bmatrix}$
b. $\begin{bmatrix} 0 & -3 \\ 3 & 0 \end{bmatrix}$
c. $\begin{bmatrix} 0 & 3 \\ -3 & 0 \end{bmatrix}$
d. $\begin{bmatrix} -3 & 0 \\ 0 & -3 \end{bmatrix}$

23. What is the maximum value of the objective function $P = 100x + 200y$ in the feasible region shown in the graph?
   a. 2800　　　 b. 2000　　　 c. 1800　　　 d. 2600

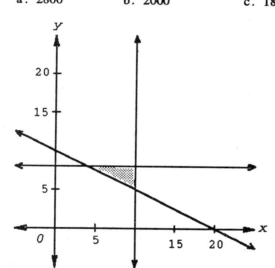

24. What is the minimum value of the function $C = x + 3y$ in the feasible region shown in the graph?
    a. 22         b. 16         c. 82         d. 28

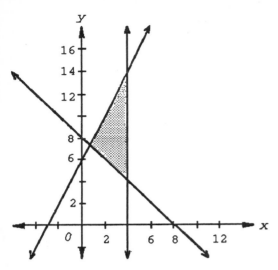

*SHORT ANSWER   Write the answer in the space provided.*

25. Create a matrix $M$ with the following elements: $m_{1\,1}$ is 3, $m_{1\,2}$ is $\pi$, $m_{13}$ is $-\pi$, $m_{21}$ is $\sqrt{2}$, $m_{22}$ is $-4.2$, and $m_{23}$ is $\frac{4}{5}$.

_____

26. Use matrices to represent the following set of linear equations.
$$\frac{3}{4}x - \frac{1}{2}y = \frac{2}{3}$$
$$2x + \frac{4}{3}y = -1$$

_____

27.
Let $A = \begin{bmatrix} 7 & 0 \\ -2 & 1 \\ 3 & -4 \end{bmatrix}$ and $B = \begin{bmatrix} -8 & -1 \\ 3 & 1 \\ -7 & 0 \end{bmatrix}$. Find $A - B$.

_____

28.
Let $J = \begin{bmatrix} 7 & 6 \\ 0 & 5 \end{bmatrix}$ and $K = \begin{bmatrix} 0 & 5 \\ 7 & 6 \end{bmatrix}$. Find $JK$ and $KJ$.

_____

29.
Let $A = \begin{bmatrix} 1 & -1 \\ 0 & -1 \end{bmatrix}$ and let $B = \begin{bmatrix} 2 & 0 \\ 0 & -7 \end{bmatrix}$. Find $(AB)A$.

_____

30. Determine whether the system of equations is inconsistent, dependent, or independent.
$$\begin{cases} 5x - 3y = 17 \\ x - 4y = 19 \end{cases}$$

_____

31. Determine whether the system of equations is inconsistent, dependent, or independent.
$$\begin{cases} -\frac{2}{3}x + 4y = 6 \\ 2x + 12y = 18 \end{cases}$$

_____

32. Determine whether the system of equations is inconsistent, dependent, or independent. How many solutions does the system have?
$$\begin{cases} -\frac{4}{5}x + \frac{9}{10}y = 0 \\ 8x - 9y = 17 \end{cases}$$

_____

33. Solve the system of equations using elimination.
$$\begin{cases} 5x - 7y = 1 \\ 4x - 6y = 2 \end{cases}$$

_____

34. Solve the system of equations using elimination.
$$\begin{cases} 8x + 9y = 10 \\ 10x + 6y = 9 \end{cases}$$

_____

35. Write the augmented matrix for the system of equations.
$$\begin{cases} 2.8x - 1.7y + 0.8z = 25.32 \\ 0.9x + 11.2y + 5.3z = 17.04 \\ -4.7x + 2.6y - 1.3z = 12.21 \end{cases}$$

_____

36.

Let $D = \begin{bmatrix} 0 & -1 & 2 & : & 11 \\ 6 & 0 & 1 & : & -3 \\ 5 & 2 & 0 & : & 2 \end{bmatrix}$. Perform the row operation $2R_1 + R_2 \rightarrow$

$R_2$ on matrix $D$.

---

37.

Let $R = \begin{bmatrix} 9 & 2 & -6 & : & 12 \\ 3 & 0 & 1 & : & -5 \\ -1 & 7 & 11 & : & 20 \end{bmatrix}$. Perform the matrix operation $R_1 + 8R_3 \rightarrow$

$R_1$ on matrix $R$.

---

38. Solve the system of equations using row reduction and back substitution.
$\begin{cases} 5x + 3y = 21 \\ 4x + 4y = 20 \end{cases}$

---

39. Solve the system of equations using row reduction and back substitution.
$\begin{cases} 5x - 3y = 16 \\ 2x + 3y = 2z = 33 \\ y + z = 10 \end{cases}$

---

40. Use your graphics calculator to find the inverse of matrix $F$, if it exists. Round numbers to the nearest hundredth. If $F^{-1}$ does not exist, write "does not exist."

$F = \begin{bmatrix} 1 & -1 & 2 \\ -2 & 1 & 3 \\ 1 & 0 & -1 \end{bmatrix}$

---

41. Use your graphics calculator to find the inverse of matrix $Q$, if it exists. Round numbers to the nearest hundredth. If $Q^{-1}$ does not exist, write "does not exist."

$$Q = \begin{bmatrix} -1 & 10 & 6 \\ 4 & 9 & -2 \\ \frac{1}{2} & -5 & -3 \end{bmatrix}$$

---

42. Use matrix algebra and your graphics calculator to solve the following system of equations.
$$\begin{cases} -9x + y = 5z \\ x + 11y = 3z - 4 \\ 8x + 14y + z = 25 \end{cases}$$

---

43. Use matrix algebra and your graphics calculator to solve the following system of equations.
$$\begin{cases} 6x - 2y + z = 76 \\ 2x + y - \frac{1}{2}z = 27 \\ -12x + 3z = 30 \end{cases}$$

---

44. Let $R = \begin{bmatrix} 1 & 1 & -1 & -1 \\ 4 & -4 & -4 & -4 \end{bmatrix}$. Write a matrix that can be used to enlarge the object represented by $R$ to twice its size.

---

45. Let $X = \begin{bmatrix} 3 & 5 & 1 & -1 \\ 3 & 3 & -1 & -1 \end{bmatrix}$. Plot the vertices of the polygon represented by $X$. Connect the vertices to form the polygon.

46. Graph the solution to the following system of inequalities:
$$\begin{cases} x + 3y \geq 9 \\ x \leq 15 - 3y \\ x \geq 0 \\ y \geq 2 \end{cases}$$

47. Graph the solution to the following system of inequalities:
$$\begin{cases} x \geq 2 \\ y \geq 3 \\ x \leq 6 \\ y \leq 8 \end{cases}$$

48. Graph the solution to the following system of inequalities:
$$\begin{cases} x + y \leq 5 \\ x \leq y \\ x \geq 0 \end{cases}$$

49. Graph the solution to the following system of inequalities:

$$\begin{cases} x \geq y \\ x \leq 2y + 2 \\ y \geq 0 \\ x \leq 5 \end{cases}$$

50. Luis works at least 18 hours per week. He studies for at least 12 hours per week. He spends 40 hours or less per week on work and study. Let $x$ be the number of hours Luis works per week, and let $y$ be the number of hours he studies per week. Write a system of three linear inequalities to describe Luis's situation.

---

51. Graph the system of inequalities you wrote to describe Luis's situation.

52. Graph the feasible region for the constraints.

$$\begin{cases} x + y \geq 5 \\ y \geq x - 5 \\ y \leq 5 \\ x \leq 8 \end{cases}$$

53. Graph the feasible region for the constraints.

$$\begin{cases} 0 \leq -y + 5 \\ 0 \leq x \leq 4 \\ x + y \geq 1 \end{cases}$$

54. Graph the feasible region for the constraints.

$$\begin{cases} 3y \leq x + 6 \\ x \leq 3y + 3 \\ y + 2x \geq 4 \\ y \leq 8 - 2x \end{cases}$$

55. Graph the feasible region for the constraints.

$$\begin{cases} y \leq 2x - 3 \\ y \geq 10 - 2x \\ y \geq 1 \end{cases}$$

56. Graph the feasible region for the constraints.

$$\begin{cases} x \leq 10 \\ x \geq 2 \\ 1 \leq y \\ 5 \geq y \\ x + 2y \leq 16 \end{cases}$$

57. Graph the feasible region for the constraints.

$$\begin{cases} y \leq x + 1 \\ y \geq 3 \\ y \leq 7 \\ x + y \leq 15 \end{cases}$$

58. What is the maximum value of the function $M = x - y$ in the feasible region shown in the graph?

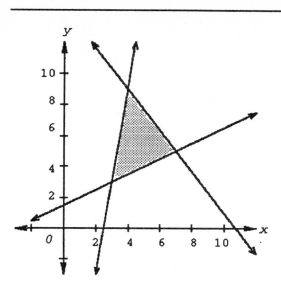

59. What is the minimum value of the function $Q = 13x - 3y$ in the feasible region shown in the graph?

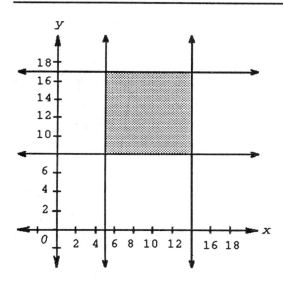

60. What is the maximum value of the function $K = 4x + 11y$ in the feasible region shown in the graph?

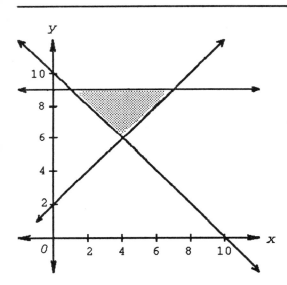

# *Answers to Chapter Questions*

1. Answer: b   Objective: 1A

2. Answer: c   Objective: 1a

3. Answer: A   Objective: 10a

4. Answer: b. 3     Objective: 1a

5. Answer:

   d. $\begin{bmatrix} 6 & 26 \\ -1 & -8 \end{bmatrix}$

   Objective: 1b

6. Answer: d. *BA* and *CB*   Objective: 2a

7. Answer:

   a. $\begin{bmatrix} -3 & 2 & 15 & 1 \\ 27 & 0 & -45 & 9 \\ 12 & 4 & 0 & 8 \end{bmatrix}$

   Objective: 2a

8. Answer: d. $179.35   Objective: 2a

9. Answer: d. all of these   Objective: 2a

10. Answer: d. 12 lbs of chocolate chip cookies and 18 lbs of double fudge cookies

    Objective: 3b

11. Answer:

    c. $\begin{bmatrix} 5 & 1 & -1 & : & -12 \\ 1 & 3 & -9 & : & 0 \\ 2 & 0 & -1 & : & -7 \end{bmatrix}$

    Objective: 4a

12. Answer:

    a. $\begin{bmatrix} -1 & 0 \\ \dfrac{2}{3} & \dfrac{1}{3} \end{bmatrix}$

    Objective: 5a

13. Answer:

d. $\begin{bmatrix} \dfrac{1}{2} & -1 \\ -\dfrac{1}{2} & 2 \end{bmatrix}$

Objective: 5a

14. Answer: d. The matrix $V$ has no inverse.  Objective: 5a

15. Answer: d. III only  Objective: 5a

16. Answer:

d. $\begin{bmatrix} 1 & -3 & 0 \\ 11.2 & 0 & 1 \\ 9 & -5 & 0 \end{bmatrix} \begin{bmatrix} x \\ y \\ z \end{bmatrix} = \begin{bmatrix} 27 \\ 4 \\ 49 \end{bmatrix}$

Objective: 6a

17. Answer:

d. $\begin{bmatrix} x \\ y \\ z \end{bmatrix} = \begin{bmatrix} 14 \\ 17 \\ 9 \end{bmatrix}$

Objective: 6a

18. Answer: c. 72  Objective: 6a

19. Answer: b. triangle  Objective: 7a

20. Answer:

d. $\begin{bmatrix} 0 & -1 \\ 1 & 0 \end{bmatrix} \begin{bmatrix} 0 & 0 & 2 \\ 3 & 0 & 0 \end{bmatrix} = \begin{bmatrix} -3 & 0 & 0 \\ 0 & 0 & 2 \end{bmatrix}$

Objective: 7a

21. Answer:

d. $\begin{bmatrix} 0 & \dfrac{1}{2} \\ -\dfrac{1}{2} & 0 \end{bmatrix}$

Objective: 7a

22. Answer:

d. $\begin{bmatrix} -3 & 0 \\ 0 & -3 \end{bmatrix}$

Objective: 7a

23. Answer: d. 2600  Objective: 10a

24. Answer: b. 16  Objective: 10a

25. Answer:

$$M = \begin{bmatrix} 3 & \pi & -\pi \\ \sqrt{2} & -4.2 & \dfrac{4}{5} \end{bmatrix}$$

Objective: 1a

26. Answer:

$$\begin{bmatrix} \dfrac{3}{4}x & -\dfrac{1}{2}y \\ 2x & \dfrac{4}{3}y \end{bmatrix} = \begin{bmatrix} \dfrac{2}{3} \\ -1 \end{bmatrix}$$

Objective: 1a

27. Answer:

$$\begin{bmatrix} 15 & 1 \\ -5 & 0 \\ 10 & -4 \end{bmatrix}$$

Objective: 1b

28. Answer:

$$JK = \begin{bmatrix} 42 & 71 \\ 35 & 30 \end{bmatrix}; \quad KJ = \begin{bmatrix} 0 & 25 \\ 49 & 72 \end{bmatrix}.$$

Objective: 2a

29. Answer:

$$(AB)A = \begin{bmatrix} 2 & -9 \\ 0 & -7 \end{bmatrix}$$

Objective: 2a

30. Answer: independent  Objective: 3a

31. Answer: independent  Objective: 3a

32. Answer: inconsistent; 0 solutions  Objective: 3a

33. Answer: $x = -4$, $y = -3$  Objective: 3b

34. Answer:
$x = \dfrac{1}{2}$, $y = \dfrac{2}{3}$
Objective: 3b

35. Answer:

$$\begin{bmatrix} 2.8 & -1.7 & 0.8 & : & 25.32 \\ 0.9 & 11.2 & 5.3 & : & 17.04 \\ -4.7 & 2.6 & -1.3 & : & 12.21 \end{bmatrix}$$

Objective: 4a


36. Answer:

$$\begin{bmatrix} 0 & -1 & 2 & : & 11 \\ 6 & -2 & 5 & : & 19 \\ 5 & 2 & 0 & : & 2 \end{bmatrix}$$

Objective: 4b


37. Answer:

$$\begin{bmatrix} 1 & 58 & 82 & : & 172 \\ 3 & 0 & 1 & : & -5 \\ -1 & 7 & 11 & : & 20 \end{bmatrix}$$

Objective: 4b


38. Answer: $x = 3$, $y = 2$   Objective: 4b

39. Answer: $x = 5$, $y = 3$, $z = 7$   Objective: 4b

40. Answer:

$$\begin{bmatrix} 0.25 & 0.25 & 1.25 \\ -0.25 & 0.75 & 1.75 \\ 0.25 & 0.25 & 0.25 \end{bmatrix}$$

Objective: 5a


41. Answer: does not exist   Objective: 5a

42. Answer: $x = -12$, $y = 7$, $z = 23$   Objective: 6a

43. Answer: $x = 13$, $y = 32$, $z = 62$   Objective: 6a

44. Answer:

$$\begin{bmatrix} 2 & 0 \\ 0 & 2 \end{bmatrix}$$

Objective: 7a

45. Answer:

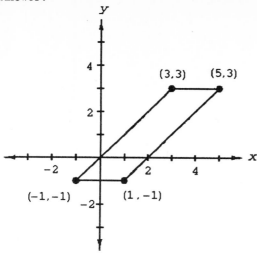

Objective: 7a

46. Answer:

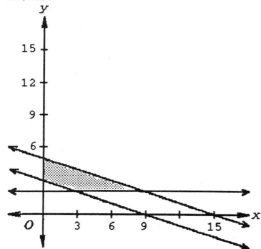

Objective: 8a

47. Answer:

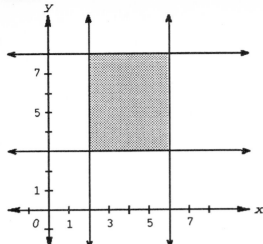

Objective: 8a

48. Answer:

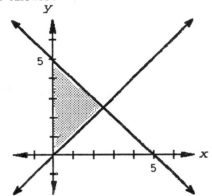

Objective: 8a

49.

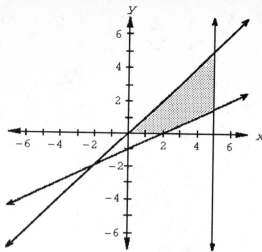

Objective: 8a

50. Answer:
$$\begin{cases} x \geq 18 \\ y \geq 12 \\ x + y \leq 40 \end{cases}$$
Objective: 8a

51.

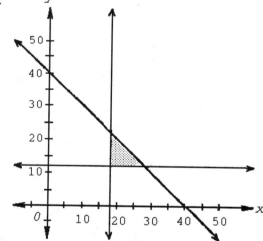

Objective: 8a

52. Answer:

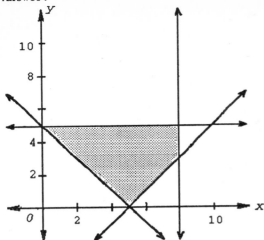

Objective: 9a


53. Answer:

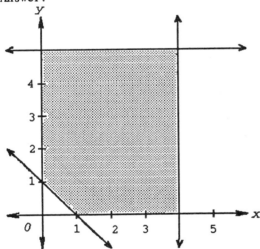

Objective: 9a

54. Answer:

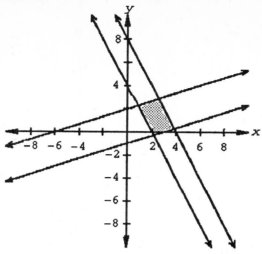

Objective: 9a


55. Answer:

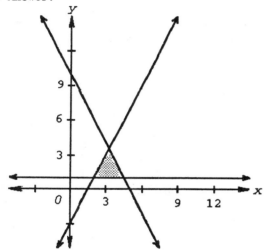

Objective: 9a

56. Answer:

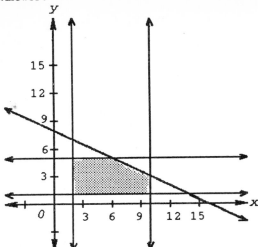

Objective: 9a

57. Answer:

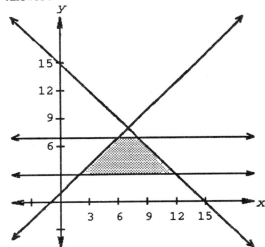

Objective: 9a

58. Answer: 2   Objective: 10a

59. Answer: 14   Objective: 10a

60. Answer: 127 at (7, 9)   Objective: 10a

# CHAPTER 5

In the space provided, write:
a. if the quantity of Column A is greater than the quantity in Column B;
b. if the quantity in Column B is greater than the quantity in Column A;
c. if the two quantities are equal; or
d. if the relationship cannot be determined from the information given.

| Column A | Column B | Answer |
|---|---|---|

1.

| The distance between points $(0, -3)$ and $(0, 5)$. | The distance between points $(-1, 2)$ and $(-4, 4)$. | _____ |
|---|---|---|

2.

| The distance between points $(8, 7)$ and $(3, -5)$. | The distance between points $(12, 2)$ and $(15, 6)$. | _____ |
|---|---|---|

3.

| number of real-number solutions of $9x^2 - 12x + 4$ | number of real-number solutions of $x^2 + x - 12$ | _____ |
|---|---|---|

*MULTIPLE CHOICE   Circle the letter of the best answer choice.*

4. A farmer wishes to fence a rectangular poultry yard. The farmer has 200 ft of fencing and wants to enclose the maximum area. What should the dimensions of the yard be so that the enclosed region has a maximum area?
   a. 20 ft by 100 ft          b. 20 ft by 20 ft
   c. 50 ft by 50 ft           d. 60 ft by 40 ft

5. Find the $x$-intercepts of the function $f(x) = -(x + 1)(x - 4)$.
   a. $(4, 0)$ $(-1, 0)$       b. $(-4, 0)$ $(0, -1)$
   c. $(4, -1)$ $(0, 0)$       d. $(4, 0)$ $(1, 0)$

6. For the function $f(x) = x^2 - x - 2$, the $x$-intercepts are:
   a. $(-1, 0)$, $(2, 0)$      b. $(0, -1)$, $(0, -2)$
   c. $(-2, 0)$, $(0, -1)$     d. $(1, 0)$, $(-2, 0)$

7. For the function $f(x) = 4 - x^2$, give the $x$-intercepts.
   a. $(0, 4)$, $(0, -4)$      b. $(2, 0)$, $(-2, 0)$
   c. $(4, 0)$, $(-4, 0)$      d. $(0, 2)$, $(0, -2)$

8. Complete the square for the quadratic expression $x^2 + 12x$.
   a. 6          b. 36          c. 24          d. 12

9. Complete the square for the quadratic expression $x^2 - 8ax$.
   a. $4a^2$          b. $8a^2$          c. $12a^2$          d. $16a^2$

10. Use the quadratic formula to solve $2x^2 - 3x = 7$. Give your answer to the nearest tenth.
    a. $(2.8, -1.3)$   b. $(2.5, -1.4)$   c. $(2.3, -1.5)$   d. $(2.9, -1.6)$

11. Use the quadratic formula to solve $2t^2 - 3t - 2 = 0$ and give your answer to the nearest tenth.
   a. (2, -0.5)   b. (-2, 0.5)   c. (0.5, 2)   d. (-2, -0.5)

12. Use the quadratic formula to solve $2x^2 = 5$ and give your answer to the nearest tenth.
   a. (4.1, -4.1)   b. (5.2, -5.2)   c. (8.3, -8.3)   d. (-1.6, 1.6)

13. Without solving, determine how many real-number solutions $x^2 + 5x + 8 = 0$ has.
   a. 2   b. 1   c. 0   d. infinitely many

14. Simplify $\sqrt{-16}$.
   a. $i$   b. $4i$   c. $2i$   d. $8i$

15. Simplify $\sqrt{2} \cdot \sqrt{-3}$.
   a. $-6i$   b. $\sqrt{6}i$   c. $6 + i$   d. $-\sqrt{6}$

16. Solve $x^2 = -9$.
   a. $3i, -3i$   b. 3, -3   c. -3   d. 3

17. Solve $r^2 = -169$.
   a. $13i$   b. $-13i$   c. -13, 13   d. $-13i, 13i$

18. Find the complex conjugate of $-4 + 8i$.
   a. $-4 - 8i$   b. $4 + 8i$   c. $-4 + 8i$   d. $8i$

19. Write the solution of $x^2 + x + 1 = 0$.
   a. $\dfrac{1 \pm i\sqrt{3}}{2}$   b. $\dfrac{1 \pm i\sqrt{2}}{3}$   c. $\dfrac{-1 \pm i\sqrt{2}}{3}$   d. $\dfrac{-1 \pm i\sqrt{3}}{2}$

20. Find the quadratic function that fits the data points (1, 4), (-1, -2), (2, 13).
   a. $f(x) = 2x^2 + 3x - 1$   b. $f(x) = 3x^2 + 2x + 1$
   c. $f(x) = 4x^2 - 2x + 1$   d. $f(x) = -4x^2 + 2x - 13$

21. Find the quadratic function that fits the data points (1, 5), (2, 9), (3, 7).
   a. $f(x) = -3x^2 + 13x - 5$   b. $f(x) = 5x^2 - 9x + 7$
   c. $f(x) = 7x^2 - x + 3$   d. $f(x) = 9x^2 - 5x + 3$

22. Which of the following is the solution set of the inequality $-x^2 + 9x - 20 < 0$?

   a.

   b.

   c.

   d.

*SHORT ANSWER   Write the answer in the space provided.*

23. A carpenter is building a rectangular shop with a fixed perimeter of 76 ft. What dimensions should the shop have in order to have the largest possible area?

_____

24. Write the function $f(x) = (x + 2)(x + 3)$ in the form $f(x) = ax^2 + bx + c$ and identify $a$, $b$, and $c$.

_____

25. Find the vertex of the graph of the function $f(x) = (x - 6)(x + 2)$.

_____

26. Find the maximum or minimum value and state the number of $x$-intercepts of the graph of the function $f(x) = 2x(3 - 2x)$.

_____

27. Use your graphics calculator to graph the function $f(x) = (2 - x)(2 + x)$, finding the minimum or maximum value, the vertex, $y$-intercept, and $x$-intercepts of the graph.

_____

28. Let $f(x) = (2x + 3)(x - 1)$. Find the vertex of the graph and determine whether it is the minimum or maximum value.

_____

29. Solve and check: $x^2 = 225$

_____

30. Solve and check: $3(x - 4)^2 = 27$

_____

31. Solve and check: $2(x^2 + 3) = 24$

_____

32. Solve and check: $2(x - 3)^2 = 18$

_____

33. Determine the distance between points $(-9, 22)$ and $(11, 1)$.

_____

34. Sketch the graph of $y = x^2$ and $y = (x - 1)^2$. What transformation is applied to $y = x^2$?

_____

35. Graph the function $f(x) = -4x^2$ and give the coordinates of the vertex, axis of symmetry, and the direction of the opening for the graph.

_____

36. Find the coordinates of the vertex and the equation of the axis of symmetry for the function $f(x) = (x - 3)^2 + 1$.

_____

37. Determine whether the function $f(x) = \frac{1}{2}(x - 1)^2 + 5$ is linear, quadratic, or neither.

_____

38. Graph the functions $f(x) = (x - 3)^2$ and $f(x) = -(x + 4)^2$. What does the coefficient in front of $x^2$ tell you about the graph?

_____

39. Draw a tile model to represent the graphic expression $x^2 + 4x + 1$.

40. Draw a tile model to represent the quadratic expression $x^2 + 6x$.

41. Solve by completing the square of $3x^2 - 21 = 2x$.

_____

42. Solve by completing the square of $5x^2 - 3x = 1$ and give your answer to the nearest tenth.

_____

43. If a playground measures 60 yards by 40 yards and a contractor wants to double it in area by extending each side an equal amount, by how much should each side be extended?

_____

44. A business owner finds that her profit $P(x)$ is a quadratic function of the number $x$ of units she sells, and that the profit function, in dollars, is given by $P(x) = 240x - x^2$. Find the number of units that yields a maximum profit, and the maximum profit. Sketch a graph of the profit function.

_____

45. Identify the real part and the imaginary part for the complex number of $5 - 12i$.

_____

46. Plot the complex number $3 + 2i$ on a complex plane.

47. Plot the complex number $-4 + 5i$ on a complex plane.

48. The cost model is $c(x) = 4x + 60$ and the revenue model is $r(x) = -x^2 + 18x$. Find the break-even points, where cost equals revenue. Use your graphics calculator to check your answer.

_____

49. A company earns $38.00 in one week, $66.00 in the second week, and $86.00 in the third week. Use the data points in the table to create a quadratic function that describes the earnings of the company as a function of weeks.

| Weeks | Earnings ($) |
| --- | --- |
| 1 | 38 |
| 2 | 66 |
| 3 | 86 |

_____

50. Use a graphics calculator to find the quadratic function of the data points (20.34, -5.86), (34.67, -6.02), (28.55, -8.46).

_____

51. The data in the table represents the accident reports in a town. Assume that a quadratic function will fit, and find the number of accidents as a function of age.

| Age of Driver | Number of Accidents |
|---|---|
| 20 | 250 |
| 40 | 150 |
| 60 | 200 |

_____

52. Find the quadratic function that fits the data (1, 0), (-1, 4), and (2, 1).

_____

53. Solve the inequality $x^2 - 2x - 15 > 0$ and graph the solutions on the number line.

_____

54. Solve the inequality $x^2 + 6x + 5 > 0$. Graph the solutions on the number line.

_____

55. If the floor area of a classroom is to be at least 200 square ft and its length is 3 ft more than twice its width, describe its width to the nearest tenth of a ft.

_____

56. Find the solution set of the inequality $x^2 - 4x > 5x - 20$ and graph on the number line.

_____

57. Find the solution set of the inequality $x^2 - 36 < 0$ and graph on the number line.

_____

# Answers to Chapter Questions

1. Answer: a   Objective: 2B

2. Answer: a   Objective: 2b

3. Answer: b   Objective: 6a

4. Answer: c. 50 ft by 50 ft   Objective: 1a

5. Answer: a. (4, 0) (-1, 0)   Objective: 1b

6. Answer: a. (-1, 0), (2, 0)   Objective: 3b

7. Answer: b. (2, 0), (-2, 0)   Objective: 3b

8. Answer: b. 36   Objective: 4b

9. Answer: d. $16a^2$   Objective: 4b

10. Answer: a. (2.8, -1.3)   Objective: 5a

11. Answer: a. (2, -0.5)   Objective: 5a

12. Answer: d. (-1.6, 1.6)   Objective: 5a

13. Answer: c. 0   Objective: 6b

14. Answer: b. $4i$   Objective: 6b

15. Answer:

    b. $\sqrt{6}i$

    Objective: 6b


16. Answer: a. $3i$, $-3i$   Objective: 6b

17. Answer: d. $-13i$, $13i$   Objective: 6b

18. Answer: a. $-4 - 8i$   Objective: 7a

19. Answer:

    d. $\dfrac{-1 \pm i\sqrt{3}}{2}$

    Objective: 7b


20. Answer: a. $f(x) = 2x^2 + 3x - 1$   Objective: 8a

21. Answer: a. $f(x) = -3x^2 + 13x - 5$   Objective: 8a

22. Answer:
    d.

Objective: 9a

23. Answer: 19 ft x 19 ft   Objective: 1a

24. Answer: $f(x) = x^2 + 5x + 6$; $a = 1$, $b = 5$, $c = 6$   Objective: 1a

25. Answer: (2, -16)   Objective: 1b

26. Answer: maximum value = 2.25; 2 $x$-intercepts   Objective: 1b

27. Answer: maximum value = 4; vertex (0, 4); $y$-intercept (0, 4); $x$-intercepts are (-2, 0), (2, 0)

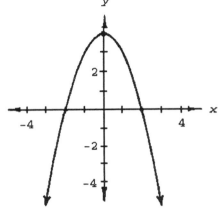

Objective: 1b

28. Answer:
    vertex $\left[-\dfrac{1}{4}, -\dfrac{25}{8}\right]$; $-\dfrac{25}{8}$ is the minimum value

Objective: 1b

29. Answer: $x = \pm 15$   Objective: 2a

30. Answer: $x = 7$ or $x = 1$   Objective: 2a

31. Answer: $x = \pm 3$   Objective: 2a

32. Answer: $x = 6$ or $x = 0$   Objective: 2a

33. Answer: 29   Objective: 2b

34. Answer: The -1 moves the graph 1 unit to the right.

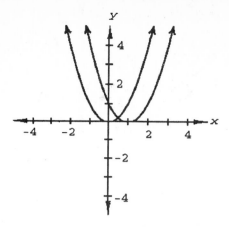

Objective: 3a

35. Answer: vertex (0,0); axis of symmetry is $x = 0$; graph opens downward

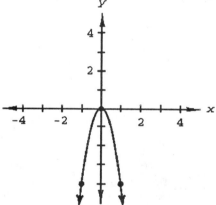

Objective: 3b

36. Answer: (3, 1); $x = 3$   Objective: 3b

37. Answer: quadratic   Objective: 3b

38. Answer: It shows the direction of the parabola.

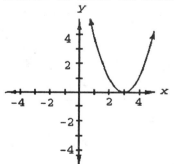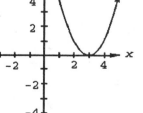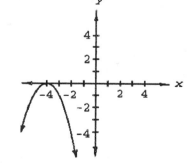

Objective: 3b

39. Answer:

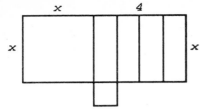

Objective: 4a

40. Answer:

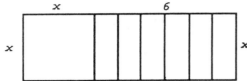

Objective: 4a

41. Answer:

$x = -2\frac{1}{3}$ or $x = 3$

Objective: 4b

42. Answer: $x = .8$ or $x = -.2$   Objective: 4b

43. Answer: 20 yards   Objective: 5a

44. Answer: Maximum profit occurs when 120 units are sold and maximum profit is $14,400.00.

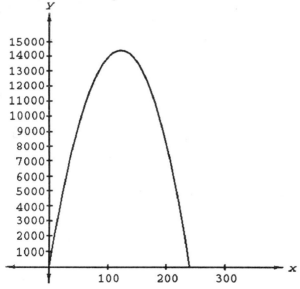

Objective: 5b

45. Answer: 5 is the real part; -12*i* is the imaginary part   Objective: 7a

46. Answer:

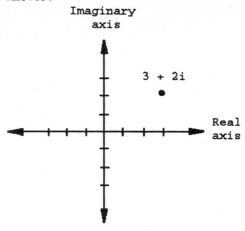

Objective: 7a

47. Answer:

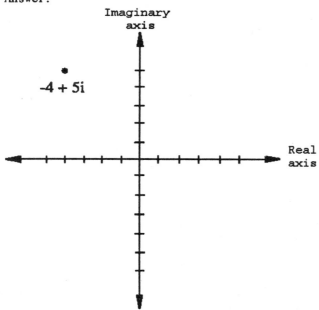

Objective: 7a

48. Answer:

$x = 7 \pm i\sqrt{11}$ There are no break-even points.
(The graphs do not intersect because the solutions are not real numbers.)

Objective: 7b

49. Answer: $f(x) = -4x^2 + 40x + 2$   Objective: 8a

50. Answer: $f(x) = 0.0499218x^2 - 2.7573651x + 29.571367$   Objective: 8a

51. Answer: $f(x) = 0.1875x^2 - 16.25x + 500$   Objective: 8a

52. Answer: $f(x) = x^2 - 2x + 1$   Objective: 8a

53. Answer: $(x > 5)$ or $(x < -3)$

Objective: 9a

54. Answer: $(x < -5)$ or $(x > -1)$

Objective: 9a

55. Answer: at least 9.3 ft   Objective: 9a

56. Answer: $(x > 5)$ or $(x < 4)$

Objective: 9a

57. Answer: $(x > -6)$ and $(x < 6)$

Objective: 9a

# CHAPTER 6

## QUANTITATIVE COMPARISON

In the space provided, write:
a. if the quantity of Column A is greater than the quantity in Column B;
b. if the quantity in Column B is greater than the quantity in Column A;
c. if the two quantities are equal; or
d. if the relationship cannot be determined from the information given.

| Column A | Column B | Answer |
|---|---|---|

1.
| degree of the polynomial function $f(x) =$ $-11x^4 - x^3 - 1x^2 + 3x - 9$ | degree of the polynomial function $f(x) =$ $x^3 + 2x^2 + 4x^6 + 2$ | _____ |
|---|---|---|

2.
| degree of the function $g(x) = 9(x + 4)^2(x - 6)$ | degree of the function $h(x) = x^3 + 4x^2$ | _____ |
|---|---|---|

## MULTIPLE CHOICE  *Circle the letter of the best answer choice.*

3. Determine the zeros for the polynomial function $f(x) = x^2 - 4x - 21$.
   a. 3, -7         b. 7, -3         c. 4, 21         d. -4, -21

4. Choose the correct expanded form of the polynomial function $f(x) = (x + 1)(x - 1)$.
   a. $f(x) = x^2 + 2$          b. $f(x) = x^2$
   c. $f(x) = 1 + x^2$          d. $f(x) = x^2 - 1$

5. Choose the correct factored form of the polynomial function $f(x) = x^4 + 3x^2 - 4$.
   a. $f(x) = (x^2 + 4)(x + 1)(x - 1)$    b. $f(x) = (x^2 + 4)(x^2 + 1)$
   c. $f(x) = (x^2 + 16)$                  d. none of these

6. Write the polynomial expression $9x^2 - 25y^2$ in factored form.
   a. $(3x + 5y)(3x - 5y)$          b. $(3x - 5y)(3x - 5y)$
   c. $(3x + 5y)(3x + 5y)$          d. $(3x + 5y)$

7. Write the polynomial expression $x^2 - 2x - 8$ in factored form.
   a. $(x + 4)(x + 2)$          b. $(x - 4)(x + 2)$
   c. $(x - 4)(x - 2)$          d. none of these

8. Write the polynomial expression $x^2 - 3x - 4$ in factored form.
   a. $(x - 4)(x + 1)$          b. $(x + 4)(x - 1)$
   c. $(x + 4)(x + 1)$          d. $(x - 4)(x - 1)$

9. Write the polynomial expression $x^2 - y^2$ in factored form.
   a. $(x + y)(x - y)$          b. $(x - y)(x - y)^2$
   c. $(x + y)(x + y)$          d. none of these

10. Write the polynomial expression $4m^2 - 2m^3 - 8m$ in factored form.
    a. $4m(2m + m - 4)$          b. $2m(2m - m^2 - 4)$
    c. $2m(m^2 + 4m - 2)$          d. none of these

11. Factor the function $f(x) = x^3 - x^2 - 14x + 24$ and find the zeros.
    a. 3, -5        b. 2, 3, -4        c. 1, -3, 4        d. -2, -3, 4

12. Find a polynomial $f(x)$ which has 3 and -5 as zeros.
    a. $f(x) = x^2 + 2x - 15$        b. $f(x) = x^2 - 2x + 15$
    c. $f(x) = x^2 - 2x - 15$        d. none of these

13. If $\frac{1}{2}$ is a zero of $f(x) = 2x^4 + x^3 - 3x^2 - x + 1$, then find 3 more
    zeros.
    a. -2, -1, 1    b. -3, 2, 1        c. -1, -1, 1        d. 3, -2, 1

14. If $-3i$ is a zero of $f(x) = x^2 + 9$, find another zero.
    a. $-3i$        b. 3        c. $3i$        d. -3

15. Find all the solutions of $6x^3 - 2x^2 - 9x + 3 = 0$.
    a. $\frac{1}{3}, \pm\sqrt{1.5}$    b. $-\frac{1}{3}, \pm\sqrt{1.5}$    c. $\frac{1}{9}, \pm\sqrt{3}$    d. $-\frac{1}{9}, \pm\sqrt{3}$

16. Complete the sentence. The only real zero of the function $f(x) = x^3 + 2x + 1$ is _____.
    a. between 1 and 2        b. between -1 and 0
    c. between -1 and 1        d. between 0 and 1

17. Find all the zeros of $f(x) = x^4 - x^3 - 3x^2 + x + 2$.
    a. -1, 1, 1, 2        b. -1, -1, 1, 2
    c. -1, 1, 1, -2        d. 1, 1, -1, 2

**SHORT ANSWER**   *Write the answer in the space provided.*

18. The perimeter of the base of a crate cannot exceed 12 ft. The height of the crate must be 2 ft less than the width of the crate. Write a function for the volume of the crate in terms of its width. Which values of the width give realistic volumes? Estimate the maximum volume of the crate.

_____

19. Write the volume of the box as a polynomial function.

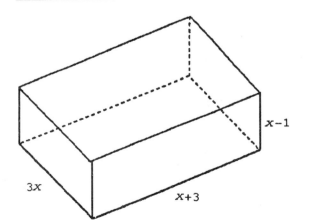

20. Use a graphics calculator to graph the polynomial function $f(x) = x^3$.
    Sketch the graph.

82

21. Write the expanded form of the polynomial function $f(x) = (x - 2)(x + 1)(x - 3)$.

_____

22. Write the factored form of the polynomial function $f(x) = x^3 - 3x^2 + 2x$.

_____

23. Find the zeros of the function $f(x) = x^3 + 4x^2 + x - 6$.

_____

24. Find the zeros of the function $f(x) = x^3 - 6x^2 + 3x + 10$.

_____

25. Factor the function $f(x) = x^4 - x^3 - 19x^2 + 49x - 30$ and find the zeros.

_____

26. Name a zero of a polynomial if $x - 3$ is a factor of a polynomial.

_____

27. Find the polynomial $f(x)$ of degree 3 with leading coefficient 1 that has -1, 1, and 2 as zeros.

_____

28. Use long division to determine if $x^2 + 9$ is a factor of $x^4 - 81$.




_____

29. Use long division to determine if $x^2 + 3x - 1$ is a factor of $x^4 - 81$.




_____

30. Use synthetic division to decide whether or not the given number $c$ is a zero of $f(x) = x^4 - x^3 - 5x^2 + 3x + 2$, $c = -2$.

_____

31. Use synthetic division to decide whether or not the given number $c$ is a zero of $f(x) = x^3 - 2x^2 + 9x - 27$, $c = 3$.

_____

32. How many crossing points and how many turning point zeros does the following function have?
$f(x) = (x + 4)^2 \ (x + 3)(x - 1)^2$

_____

33. Find a polynomial of degree 4 with 0 as a zero of multiplicity 2, and 5 as a zero of multiplicity 2.

_____

34. Find a polynomial of degree 4 with zeros at –2 and –1 and a zero of multiplicity 2 at 3.

_____

35. Is the function graphed a polynomial function of even or odd degree?

_____

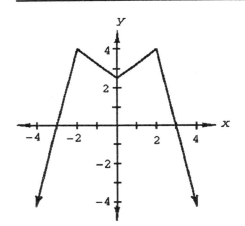

36. Is the function graphed an even or odd function?

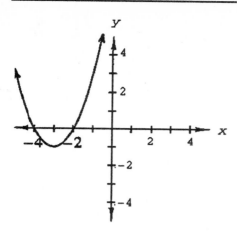

37. For the function graphed, where is the function increasing and decreasing?

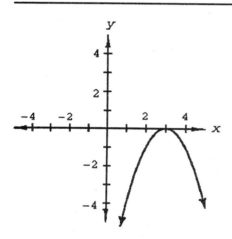

38. Sketch the graph of the polynomial function $f(x) = (x - 1)(x + 2)(x - 3)$.

39. Sketch the graph of the polynomial function $f(x) = (x - 1)^2 (x + 4)^2$.

40. June deposits $1,000, $2,000, $3,000, and $4,000 at the end of her freshman, sophomore, junior, and senior high school years respectively. At what approximate interest rate, compounded yearly, must she invest her money to cover a car purchase of $10,512 at the end of her senior year?

41. What interest rate, compounded yearly, would June need given her deposits and a car price of $10,820?

42. A block of wood has the shape of a cube. A slice 2 in. thick is cut parallel to one of the faces. What is the length of each edge of the cube if the volume of the remaining piece is 75 cubic in.?

43. A box with an open top is constructed from a rectangular piece of copper that measures 9 in. by 12 in. by cutting out from each corner a square of side $x$, and then folding up the sides. If $V$ is the volume of the box, find a formula for $V$ as a function $x$.

_____

44. Find the maximum volume of the box.

_____

45. What is the realistic domain of the volume function?

_____

46. Chris is able to deposit $300, $500, $700, and $900 at the end of his freshman, sophomore, junior, and senior years respectively. How much money will he have at the end of his senior year if the money is invested at 7% interest compounded annually? About what interest rate will Chris need if he wants to have $2675 at the end of his senior year ?

_____

47. Use variable substitution to simplify $100(1 + r)^3 + 200(1 + r)^2 + 300(1 + r) + 400 = 1062$.

_____

48. Use variable substitution to find the zeros of the following function.
$f(x)\ ) = 200(x - 1.2)^3 - 600(x - 1.2)^2 - 1200x + 1600$

_____

# Answers to Chapter Questions

1. Answer: b   Objective: 1B

2. Answer: c   Objective: 1b

3. Answer: b. 7, -3   Objective: 1b

4. Answer: d. $f(x) = x^2 - 1$   Objective: 1b

5. Answer: a. $f(x) = (x^2 + 4)(x + 1)(x - 1)$   Objective: 2a

6. Answer: a. $(3x + 5y)(3x - 5y)$   Objective: 2a

7. Answer: b. $(x - 4)(x + 2)$   Objective: 2a

8. Answer: a. $(x - 4)(x + 1)$   Objective: 2a

9. Answer: a. $(x + y)(x - y)$   Objective: 2a

10. Answer: b. $2m(2m - m^2 - 4)$   Objective: 2a

11. Answer: b. 2, 3, -4   Objective: 2b

12. Answer: a. $f(x) = x^2 + 2x - 15$   Objective: 2b

13. Answer: c. -1, -1, 1   Objective: 3b

14. Answer: c. $3i$   Objective: 3b

15. Answer:
    a. $\frac{1}{3}, \pm \sqrt{1.5}$

    Objective: 3b


16. Answer: b. between -1 and 0   Objective: 3b

17. Answer: b. -1, -1, 1, 2   Objective: 4a

18. Answer: $f(x) = x(6 - x)(x - 2)$; width must be between 2ft and 6 ft ; about 17 ft
    Objective: 1a

19. Answer: $f(x) = 3x(x + 3)(x - 1)$   Objective: 1b

20. Answer:

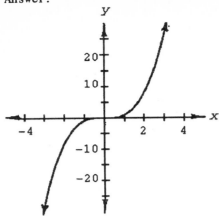

Objective: 1b


21. Answer:

$f(x) = x^3 - 4x^2 + x + 6$
Objective: 1b


22. Answer: $f(x) = x(x - 2)(x - 1)$   Objective: 2a

23. Answer: 1, -2, -3   Objective: 2b

24. Answer: 2, 5, -1   Objective: 2b

25. Answer: $P(x) = (x - 1)(x - 2)(x - 3)(x + 5)$; 1, 2, 3, -5   Objective: 2b

26. Answer: 3   Objective: 2b

27. Answer:

$f(x) = (x + 1)(x - 1)(x - 2) = x^3 - 2x^2 - x + 2$
Objective: 2b


28. Answer:
Yes, it is because the remainder is 0 and $x^4 - 81 = (x^2 - 9)(x^2 + 9)$.

$$
\begin{array}{r}
x^2 - 9 \\
x^2 + 9 \overline{)\, x^4 \qquad\quad -81} \\
\underline{x^4 + 9x^2} \\
-9x^2 - 81 \\
\underline{-9x^2 - 81} \\
0
\end{array}
$$

Objective: 3a

29. Answer:
No, it is not because the remainder is not 0 and $x^2 + 3x - 1$ is not a factor of $x^4 - 81$.

$$
\begin{array}{r}
x^2 - 3x + 10 \\
x^2 + 3x - 1 \,\overline{\smash{\big)}\, x^4 \phantom{+3x^3 - x^2 - 3x} -81} \\
\underline{x^4 + 3x^3 - x^2 \phantom{-3x - 81}} \\
-3x^3 + x^2 \phantom{-3x - 81} \\
\underline{-3x^3 - 9x^2 + 3x \phantom{- 81}} \\
10x^2 - 3x - 81 \\
\underline{10x^2 + 30x - 10} \\
-33x - 71
\end{array}
$$

Objective: 3a

30. Answer:
−2 is a zero because the remainder is 0

$$
\begin{array}{r|rrrrr}
-2 & 1 & -1 & -5 & 3 & 2 \\
   &   & -2 & 6 & -2 & -2 \\
\hline
   & 1 & -3 & 1 & 1 & 0
\end{array}
$$

Objective: 3b

31. Answer:
3 is not a zero because the remainder is not 0.

$$
\begin{array}{r|rrrr}
3 & 1 & -2 & 9 & -27 \\
  &   & 3 & 3 & 36 \\
\hline
  & 1 & 1 & 12 & 9
\end{array}
$$

Objective: 3b

32. Answer: One crossing point at −3 and two turning point zeros at −4 and 1   Objective: 4a

33. Answer: $f(x) = x^2(x - 5)^2 = x^4 - 10x^3 + 25x^2$   Objective: 4a

34. Answer: $f(x) = (x + 2)(x + 1)(x - 3)^2 = x^4 - 3x^3 - 7x^2 + 15x + 18$   Objective: 4a

35. Answer: even   Objective: 4b

36. Answer: even   Objective: 4b

37. Answer: increasing over $x$: $-\infty$ to 3; decreasing over $x$: 3 to $\infty$   Objective: 4b

38. Answer:

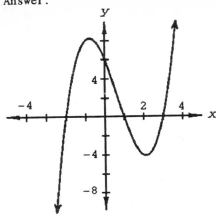

Objective: 4b


39. Answer:

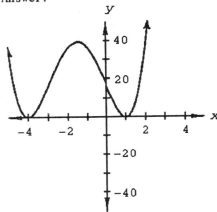

Objective: 4b


40. Answer: 5%  Objective: 5a

41. Answer: 8%  Objective: 5a

42. Answer: 5 in.  Objective: 5a

43. Answer: $V(x) = x(9 - 2x)(12 - 2x)$  Objective: 5a

44. Answer: approximately 81 in$^3$  Objective: 5a

45. Answer: $0 < x < 4.5$  Objective: 5a

46. Answer: $300(1.07)^3 + 500(1.07)^2 + 700(1.07) + 900 = \$2588.96$ ; About 10%  Objective: 5a

47. Answer: $100R^3 + 200R^2 + 300R + 400 = 1062$  Objective: 5b

48. Answer: $x = -.8;\ x = 2.2;\ x = 5.2$  Objective: 5b

# CHAPTER 7

*QUANTITATIVE COMPARISON*

In the space provided, write:
a. if the quantity of Column A is greater than the quantity in Column B;
b. if the quantity in Column B is greater than the quantity in Column A;
c. if the two quantities are equal; or
d. if the relationship cannot be determined from the information given.

| Column A | Column B | Answer |
|---|---|---|

1.

| population of Tinytown after 5 years of consistent 3% decline | population of Tinytown after 8 years of consistent 3% decline | _____ |

2.

| population of Westville after 10 years of consistent 6% decline | population of Freeburg after 8 years of consistent 6% decline | _____ |

3.

| $\left(\dfrac{1}{4}\right)^{-2}$ | $2^3$ | _____ |

4.

| $3,000 invested for 20 years at 4% interest, compounded annually | $2,700 invested for 20 years at 4% interest, compounded quarterly | _____ |

5.

| log 97 | $\sqrt{5}$ | _____ |

*MULTIPLE CHOICE  Circle the letter of the best answer choice.*

6. The population of Brontislaw was 11,020 in 1990 and was growing at a rate of 1.1% per year. Assuming that the population of Brontislaw continues to grow at the same rate, use your calculator to find Brontislaw's projected population in the year 2000.
   a. 12,294       b. 13,438       c. 11,996       d. 28,583

7. In January 1995, the 6,230 residents of Floodplains began leaving at a rate of 60% per year. If that rate stays constant, when will there be fewer than 100 residents?
   a. by 1996       b. by 2000       c. by 2004       d.  by 2008

8. Which graph shows the function $f(x) = \left[\frac{1}{2}\right]^{x+4}$?

a.

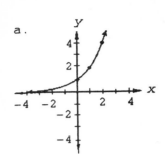

b.

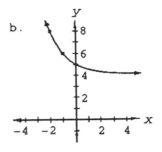

c.

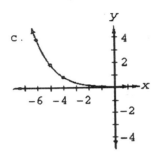

d.

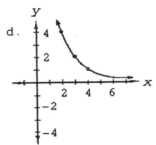

9. Suppose $4,000 is invested at 3% interest, compounded monthly. How much is the investment worth after 5 years?
a. $23,566.41    b. $6,262.72    c. $4,646.46    d. $4,060.44

10. Let $f(x) = 2x + 1$. Which graph represents $f^{-1}$?

a.

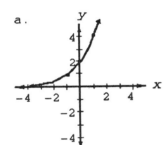

b.

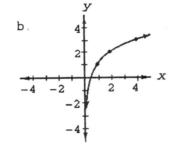

c.

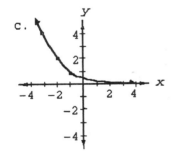

d.

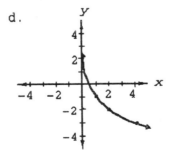

11. What logarithmic equation is equivalent to the exponential equation $3^7 = 2187$?
a. $\log_{2187} 3 = 7$   b. $\log_7 3 = 2187$    c. $\log_3 7 = 2187$   d. $\log_3 2187 = 7$

12. What exponential equation is equivalent to the logarithmic equation $\log_5 \left[\frac{1}{25}\right] = -2$?

a. $(-5)^2 = \frac{1}{25}$   b. $5^{\frac{1}{25}} = -2$    c. $5^{-2} = \frac{1}{25}$    d. $\left[\frac{1}{25}\right]^{-2} = 5$

13. Which expression is equivalent to $\log_{10} 90$?
   a. $\log_{10} 27 + \log_{10} 63$       b. $\log_{10} 15 + \log_{10} 6$
   c. $\log_{10} 120 - \log_{10} 30$       d. $\log_{10} 30 - \log_{10} 3$

14. Which expression is equivalent to $\log_3 7$?
   a. $\log_3 91 - \log_3 13$       b. $\log_3 4 + \log_3 3$
   c. $\log_3 21 - \log_3 14$       d. $\log_3 5 + \log_3 2$

15. Which expression is equivalent to $\log_5 x^3$?
   a. $\log_5 3x$       b. $3 \log_5 x$       c. $\log_5 3 + \log_5 x$   d. $\log_5 3 - \log_5 x$

16. Which single logarithm is equivalent to $\log_3 12 + \log_3 5 - \log_3 4$?

   a. $\log_3 13$       b. $\log_3 \left[\dfrac{17}{4}\right]$       c. $\log_3 15$       d. $\log_3 20$

17. Solve for $x$.
   $\log_4 (x + 1) + \log_4 (x - 2) = \log_4 (x^2 - 9)$
   a. $x = 3$       b. $x = 15$       c. $x = 10$       d. $x = 7$

18. Use your calculator to find $e^8$. Round your answer to the nearest hundredth.
   a. 22026.47       b. 2.08       c. 54.60       d. 2980.96

19. Use your calculator to find ln 123. Round your answer to the nearest hundredth.
   a. 5.77       b. 5.44       c. 4.81       d. 6.80

20. Tony invests $2,000 at 4.5% interest, compounded continuously. What is the value of his investment after 6 years? Round your answer to the nearest whole cent.
   a. $3,297.44       b. $2,619.93       c. $2,866.66       d. $29,759.46

21. How long will it take an investment of $1,750 to double in value at 4% interest compounded continuously?
   a. 29.66 years       b. 6.47 years       c. 17.33 years       d. 12.53 years

22. A sum of money is invested at a certain annual interest rate, compounded continuously. After 10.2 years, the investment has doubled. At what interest rate was the money invested?
   a. 5.7%       b. 6.8%       c. 7.6%       d. 2.9%

23. Solve by graphing:
   $\log(2x + 6) + \log(x - 2) = 2$
   a. $x = 1.57$       b. $x = 7$       c. $x = 15$       d. $x = 4$

24. Solve by graphing.
   $12e^{3x+2} = 60$.
   a. $x = 4$       b. $x = 2.5$       c. $x = -0.67$       d. $x = -0.13$

25. Solve by graphing.
   $\log(x - 6) + \log(2x - 11) = 1$
   a. $x = 5$       b. $x = 13$       c. $x = 9$       d. $x = 8$

*SHORT ANSWER*   *Write the answer in the space provided.*

26. The population of Willenburg was 5,041 in 1990 and was declining at a rate of 1.2% per year. Assuming that Willenburg's population continues to decline at the same rate, find the projected population of Willenburg in the year 2000.

27. A 470-gram sample of a radioactive substance decreases in mass by 5% each year. What is the mass of the sample after 8 years? Round your answer to the nearest gram.

28. The population of Dullsville was 7,230 in 1990 and was declining at a rate of 1.1% per year. Assuming that the population of Dullsville continues to decline at the same rate, use your calculator to find Dullsville's projected population in the year 2000.

---

29. A 800-gram sample of a radioactive substance decreases in mass by 50% each year. What is the mass of the sample after 4 years?

---

30. Graph $f(x) = 2^x$ and $g(x) = 2^x - 3$. Explain how the graph of $g(x)$ can be obtained from the graph of $f(x)$.

---

31. Graph $f(x) = 2^x$ and $h(x) = 2^{x+1}$. Explain how the graph of $h(x)$ can be obtained from the graph of $f(x)$.

---

32. Graph $g(x) = \left[\frac{1}{2}\right]^x$ and $h(x) = \left[\frac{1}{2}\right]^{x-3}$. Explain how the graph of $h(x)$ can be obtained from the graph of $g(x)$.

---

33. Graph $g(x) = \left[\frac{1}{2}\right]^x$ and $h(x) = \left[\frac{1}{2}\right]^x + 4$. Explain how the graph of $h(x)$ can be obtained from the graph of $g(x)$.

---

34. Suppose $2,500 is invested at 4% interest, compounded quarterly. How much is the investment worth after 6 years? Round your answer downward to the nearest whole cent.

---

35. Graph $f(x) = 2^x$. What function is the inverse of $f$? Sketch $f^{-1}$ on the same coordinate plane.

---

36. Graph $f(x) = 2^{-x}$; then sketch $f^{-1}$ on the same coordinate plane.

37. Write an equivalent logarithmic equation.
$7^{-2} = \dfrac{1}{49}$

---

38. Write an equivalent exponential equation.
$\log_{10}(1,000,000) = 6$

---

39. Write an equivalent logarithmic equation.
$9^3 = 729$

---

40. Write an equivalent exponential equation.
$\log_2 1024 = 10$

---

41. Express $\log_2 x(x - 3)$ as a sum of logarithms.

---

42. Find the exact value of the expression $\log_5 (25)^7 - \log_5 125$.

---

43. Solve for $x$.
$3 \log_2 4 + 4 \log_2 3 - 2 \log_2 6 = 2 \log_2 x$

---

44. Solve for $x$.
$\log_{10} (x^2 + 4) - \log_{10} (x - 6) = \log_{10} (2x + 3)$

---

45. Use your calculator to solve for $x$. Round your answer to the nearest hundredth.
$\log x = 4.52$

---

46. Use your calculator to solve for $x$. Round your answer to the nearest hundredth.
$10^x = 57$

---

47. Use your calculator to solve for x. Round your answer to the nearest hundredth.
$\log x = -0.0083$

---

48. Use your calculator to solve for $x$. Round your answer to the nearest hundredth.
$10^x = \dfrac{1}{4}$

---

49. Write an equivalent exponential equation.
$\log x = 0.73$

---

50. Write an equivalent logarithmic equation.
$10^x = 42.7$

---

51. Write an equivalent exponential equation.
$\log x = -5.5$

---

52. Write an equivalent logarithmic equation.

$10^x = 0.039$

_____

53. Write an equivalent logarithmic equation.

$e^{\sqrt{2}} = 4.11$

_____

54. Write an equivalent exponential equation.

$\ln \dfrac{3}{4} = -0.29$

_____

55. A radioactive substance decays according to the decay function $A = A_0 e^{-(.12t)}$, where $A_0$ is the amount of the substance initially present and $t$ is the time in years. How long will it take a sample of this substance to decay to $\dfrac{1}{2}$ its original amount? Round your answer to the nearest hundredth of a year.

_____

56. Solve using the properties of exponential and logarithmic functions.

$10^{6x-3} = 100,000$

_____

57. Solve using the properties of exponential and logarithmic functions.

$\ln(4x - 9) - \ln(x + 4) = \ln(2x - 39)$

_____

58. Solve using the properties of exponential and logarithmic functions.

$\log(50x - 100) = 3$

_____

59. Solve using the properties of logarithmic and exponential functions.

$e^{5x+11} = e^{6x-5}$

_____

60. Use your calculator to solve for $x$. Round your answer to the nearest hundredth.

$e^{7x-5} = 15$

_____

61. Use your calculator to solve for $x$. Round your answer to the nearest hundredth.

$7^x = 526$

_____

# *Answers to Chapter Questions*

1. Answer: a   Objective: 1A

2. Answer: d   Objective: 1a

3. Answer: a   Objective: 2a

4. Answer: a   Objective: 2b

5. Answer: b   Objective: 5a

6. Answer: a. 12,294   Objective: 1a

7. Answer: b. by 2000   Objective: 1a

8. Answer:

   c.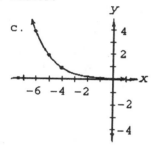

   Objective: 2a

9. Answer: c. $4,646.46   Objective: 2b

10. Answer:

    b.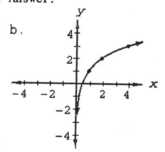

    Objective: 3a

11. Answer: d. $\log_3 2187 = 7$   Objective: 3b

12. Answer:

    c. $5^{-2} = \dfrac{1}{25}$

    Objective: 3b

13. Answer: b. $\log_{10} 15 + \log_{10} 6$  Objective: 4a

14. Answer: a. $\log_3 91 - \log_3 13$  Objective: 4a

15. Answer: b. $3 \log_5 x$  Objective: 4a

16. Answer: c. $\log_3 15$  Objective: 4b

17. Answer: d. $x = 7$  Objective: 4b

18. Answer: d. 2980.96  Objective: 6a

19. Answer: c. 4.81  Objective: 6a

20. Answer: b. $2,619.93  Objective: 6b

21. Answer: c. 17.33 years  Objective: 6b

22. Answer: b. 6.8%  Objective: 6b

23. Answer: b. $x = 7$  Objective: 7a

24. Answer: d. $x = -0.13$  Objective: 7a

25. Answer: d. $x = 8$  Objective: 7a

26. Answer: 4,468  Objective: 1a

27. Answer: 312 grams  Objective: 1a

28. Answer: 6,473  Objective: 1a

29. Answer: 50 grams  Objective: 1a

30. Answer: The graph of $g(x)$ can be obtained from the graph of $f(x)$ by moving it 3 units downward.

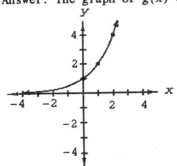 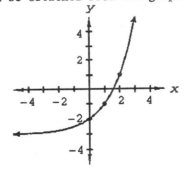

Objective: 2a

31. Answer: The graph of $h(x)$ can be obtained from the graph of $f(x)$ by moving it 1 unit to the left.

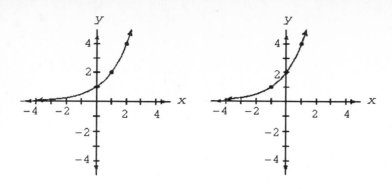

Objective: 2a

32. Answer:
The graph of $h(x)$ can be obtained from the graph of $g(x)$ by moving it 3 units to the right.

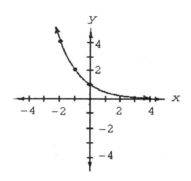
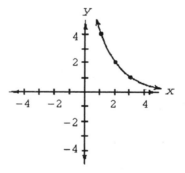

Objective: 2a

33. Answer: The graph of $h(x)$ can be obtained from the graph of $g(x)$ by moving it 4 units upward.

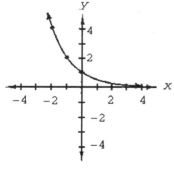
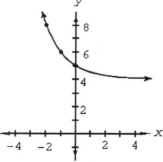

Objective: 2a

34. Answer: $3,174.33   Objective: 2b

35. Answer: $f^{-1}(x) = \log_2 x$

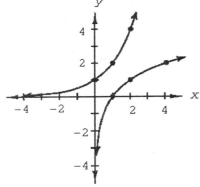

Objective: 3a

36. Answer:

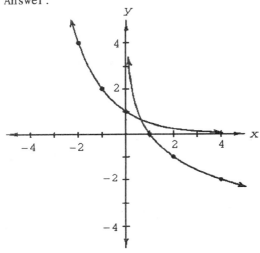

Objective: 3a

37. Answer:

$$\log_7\left[\frac{1}{49}\right] = -2$$

Objective: 3b

38. Answer: $10^6 = 1,000,000$   Objective: 3b

39. Answer: $\log_9 729 = 3$   Objective: 3b

40. Answer: $2^{10} = 1024$   Objective: 3b

41. Answer: $\log_2 x(x - 3) = \log_2 x + \log_2 (x - 3)$   Objective: 4a

42. Answer: 11   Objective: 4b

43. Answer: $x = 12$   Objective: 4b

44. Answer: $x = 11$   Objective: 4b

45. Answer: $x = 33,113.11$   Objective: 5a

46. Answer: $x = 1.76$   Objective: 5a

47. Answer: $x = 0.98$   Objective: 5a

48. Answer: $x = -0.60$   Objective: 5a

49. Answer: $x = 10^{0.73}$   Objective: 5b

50. Answer: $x = \log 42.7$   Objective: 5b

51. Answer: $x = 10^{-5.5}$   Objective: 5b

52. Answer: $x = \log 0.039$   Objective: 5b

53. Answer:

   $$\ln 4.11 = \sqrt{2}$$
   Objective: 6a


54. Answer:
   $$e^{-0.29} = \frac{3}{4}$$

   Objective: 6a


55. Answer: 5.78 years   Objective: 6b

56. Answer:
   $$x = \frac{4}{3}$$

   Objective: 7b


57. Answer:
   $$x = 21 \text{ or } x = -\frac{7}{2}$$

   Objective: 7b


58. Answer: $x = 22$   Objective: 7b

59. Answer: $x = 16$   Objective: 7b

60. Answer: $x = 1.10$   Objective: 7b

61. Answer: $x = 3.22$   Objective: 7b

# CHAPTER 8

## QUANTITATIVE COMPARISON

In the space provided, write:
a. if the quantity of Column A is greater than the quantity in Column B;
b. if the quantity in Column B is greater than the quantity in Column A;
c. if the two quantities are equal; or
d. if the relationship cannot be determined from the information given.

| Column A | Column B | Answer |
|---|---|---|

1.

| $3 \tan 45°$ | $6 \cos 60°$ | _____ |
|---|---|---|

2.

| $\frac{1}{2} \tan 60°$ | $\tan 30°$ | _____ |
|---|---|---|

3.

| the $x$-coordinate shown on the graph | the $y$-coordinate shown on the graph | _____ |
|---|---|---|

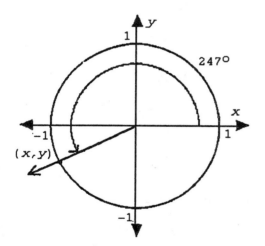

4.

| period of $f(\theta) = \frac{1}{3} \cos \theta$ | period of $g(\theta) = \cos \frac{1}{3} \theta$ | _____ |
|---|---|---|

5.

| amplitude of $f(\theta) = 2 \sin 8 \theta$ | amplitude of $g(\theta) = -2 \cos \frac{1}{8} \theta$ | _____ |
|---|---|---|

6.

| $\cos \frac{\pi}{6}$ | $\frac{\sqrt{3}}{4}$ | _____ |
|---|---|---|

*MULTIPLE CHOICE Circle the letter of the best answer choice.*

7. What trigonometric ratio describes $\dfrac{x}{\sqrt{x^2 + y^2}}$?

   a. tan 60°          b. sin 60°
   c. cos 60°          d. none of these

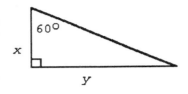

8. Find $x$.

   a. 5          b. $5\sqrt{2}$          c. $\dfrac{5\sqrt{2}}{2}$          d. $\dfrac{5\sqrt{3}}{2}$

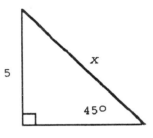

9. Find $y$.

   a. $16\sqrt{3}$          b. $\dfrac{16\sqrt{3}}{3}$          c. 32          d. 8

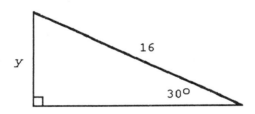

10. What is the measure of $\angle A$?
    a. 30°          b. 60°          c. 45°          d. 50°

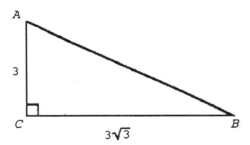

11. Find *w*.

    a. $2\sqrt{2}$          b. $4\sqrt{2}$          c. $\dfrac{8}{\sqrt{3}}$          d. $2\sqrt{3}$

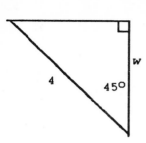

12. What are the coordinates of the point (*x*, *y*) shown in the graph?

    a. $\left[\dfrac{1}{2}, \dfrac{\sqrt{3}}{2}\right]$      b. $\left[\dfrac{\sqrt{3}}{2}, \dfrac{1}{\sqrt{2}}\right]$      c. $\left[\dfrac{1}{\sqrt{2}}, \dfrac{1}{\sqrt{2}}\right]$      d. $\left[\dfrac{\sqrt{3}}{2}, \dfrac{1}{2}\right]$

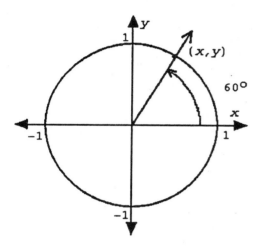

13. What are the coordinates of the point (*x*, *y*) shown in the graph?

    a. $\left[\dfrac{1}{\sqrt{2}}, \dfrac{1}{\sqrt{2}}\right]$      b. $\left[\dfrac{1}{\sqrt{2}}, -\dfrac{1}{\sqrt{2}}\right]$      c. $\left[-\dfrac{1}{\sqrt{2}}, -\dfrac{1}{\sqrt{2}}\right]$      d. $\left[-\dfrac{1}{\sqrt{2}}, \dfrac{1}{\sqrt{2}}\right]$

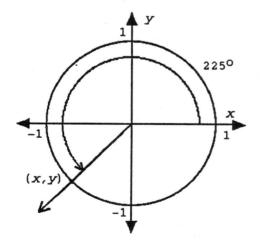

14. Use your calculator to find the coordinates of the point $(x, y)$ shown on the graph. Round both coordinates to the nearest hundredth.
    a. (-0.80, 0.60)   b. (-0.80, -0.60)
    c. (0.80, 0.60)    d. (0.80, -0.60)

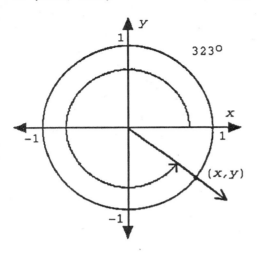

15. Which pair of angles is coterminal?
    a. 45° and -45°    b. 135° and -45°
    c. -225° and 45°   d. -315° and 45°

16. Find m∠HJK in △HJK. Round your answer to the nearest hundredth of a degree.
    a. 0.38°       b. 38.46°       c. 22.62°       d. 67.38°

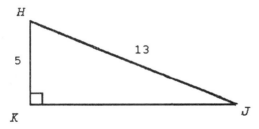

17. Find the length of side $\overline{DE}$ of △DEF. Round your answer to the nearest hundredth of a unit.
    a. 17.01 m     b. 58.77 m     c. 0.58 m     d. 12.36 m

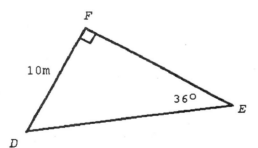

106

18. Use inverse trigonometric functions to find ∠*DOE*. ∠*DOE* is in standard position.
    a. 45°          b. 135°          c. –135°          d. 315°

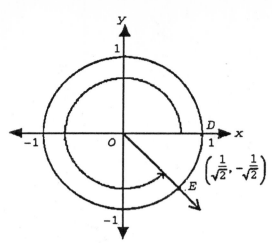

19. Which function is shown in the graph?

    a. $f(\theta) = \dfrac{4}{3} \sin 2\theta$          b. $g(\theta) = \dfrac{3}{4} \cos 2\theta$

    c. $h(\theta) = 2 \sin \dfrac{3}{2}\theta$          d. $k(\theta) = 2 \cos \dfrac{3}{4}\theta$

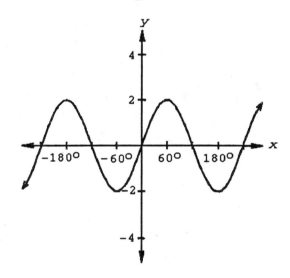

20. Which function is shown in the graph?

 a. $f(\theta) = \sin \frac{1}{2}(\theta + 45)$    b. $g(\theta) = \frac{1}{2}\sin(\theta - 45)$

 c. $h(\theta) = \sin \left[\frac{1}{2}\theta + 45\right]$    d. $k(\theta) = \sin 2(\theta - 45)$

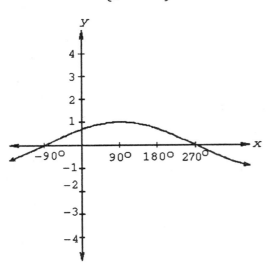

21. What are the vertical shift and phase shift of the function

 $f(\theta) = \frac{3}{4}\cos 2(\theta + 30) + 1$?

 a. vertical shift -1, phase shift 60
 b. vertical shift 1, phase shift -30
 c. vertical shift -1, phase shift 30
 d. vertical shift 1, phase shift -60

22. Find the length of $\overset{\frown}{QRS}$. Round your answer to the nearest hundredth of a foot.
 a. 8250 ft    b. 45.83 ft    c. 143.99 ft    d. none of these

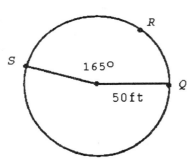

23. Find the area of sector WXZ. Round your answer to the nearest hundredth of a square inch.
 a. 1.95 in$^2$    b. 47.89 in$^2$    c. 683 in$^2$    d. 95.55 in$^2$

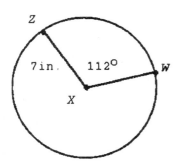

24. Find the central angle that intercepts an arc length of 12 in. on a circle with a radius of 14 in.

   a. 1.17          b. 1.05          c. .86          d. $\dfrac{\pi}{2}$

25. Find the central angle for a sector with an area of 100 cm$^2$ in a circle with a radius of 8 cm.

   a. 179.05°        b. 3.13°          c. 31.25°        d. 154.60°

26. The average number of inches of rainfall in the $x$th month of the year in Mimosa City is modeled by the function $R(x) = 4 + 3.5 \cos \dfrac{\pi}{6}$

   $(x - 1)$. During what month is the average rainfall in Mimosa City a minimum?

   a. January       b. July          c. June          d. February

27. Find two months during which the average rainfall in Mimosa City is about 4 in.

   a. April and October          b. April and July
   c. October and November       d. none of these

*SHORT ANSWER*   *Write the answer in the space provided.*

28. Find $x$.

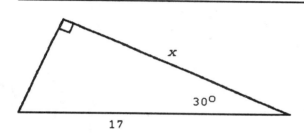

29. Find $y$.

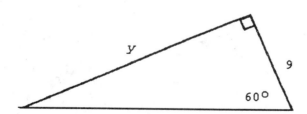

30. What are the coordinates of the point (x, y) shown on the graph?

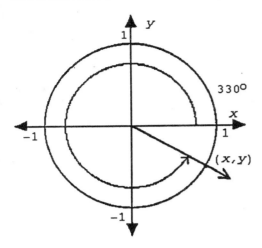

31. What are the coordinates of the point (x, y) shown on the graph?

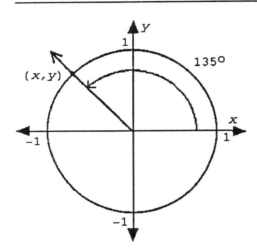

32. Use your calculator to find the coordinates of the point (x, y) shown on the graph. Round both coordinates to the nearest hundredth.

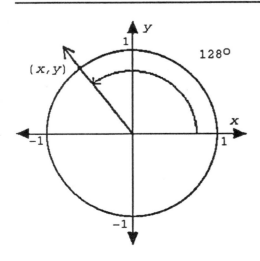

33. Find a coterminal angle for 760°

34. Use your calculator to find ∠D in ΔDEF. Round your answer to the nearest hundredth of a degree.

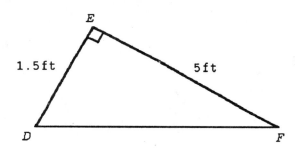

35. Use your calculator to find side $\overline{AB}$ of ΔABC. Round your answer to the nearest hundredth of a foot.

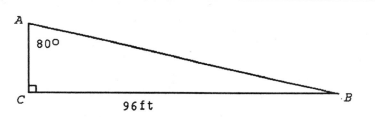

36. Find the length of side $\overline{XY}$ of ΔXYZ.

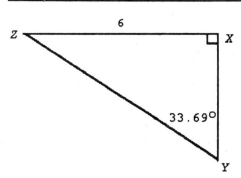

37. Use inverse trigonometric functions to find m∠POQ. ∠POQ is in standard position.

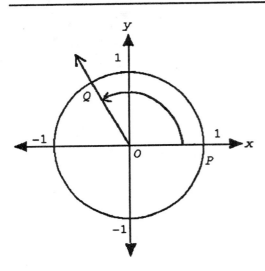

38. Which function is shown in the graph?

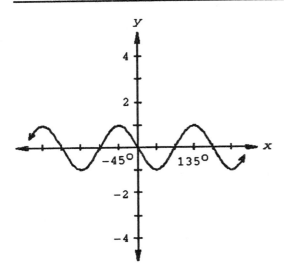

39. What is the period of the function $f(\theta) = 10 + 5 \cos 3(\theta + 30)$?

40. What is the amplitude and the period of the function $g(\theta) = \frac{1}{2} \sin 4\theta$?

41. Give the exact value of $\frac{\pi}{3}$.

42. Give the exact value of $\frac{\pi}{4}$.

_____

43. Convert 135° to radians.

_____

44. Write a function of the form $f(x) = a + b \cos c(x - d)$ for the graph.

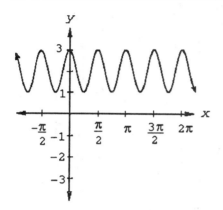

45. What is the period (in radians) of $g(x) = 17 - 3 \sin \frac{1}{3} \left[ x - \frac{\pi}{4} \right]$?

_____

46. Find the amplitude and the period of the function $f(x) = -3 \cos \frac{2}{3}x$ and graph the function.

_____

47. What is the phase shift (in radians) and the vertical shift of $h(x) = 5.3 + \sqrt{2} \sin 4 \left[ x - \frac{\pi}{3} \right]$?

_____

48. Find the length of $\overset{\frown}{ABC}$. Round your answer to the nearest hundredth of a meter.

_____

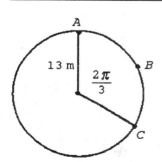

49. Find the area of sector *KOL*. Round your answer to the nearest hundredth of a square centimeter.

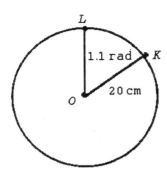

50. Find the length of $\overset{\frown}{PQ}$ and the area of sector *POQ*.

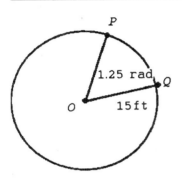

51. Find the length of $\overset{\frown}{AB}$ and the area of sector *AOB*. Round your answers to the nearest hundredth.

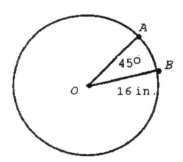

52. The temperature on a day in April in the town of Bleeville can be modeled by the function where $f(x) = 62 + 18 \sin \frac{\pi}{12} (x - 6)$ where $f(x)$ is the temperature in degrees Fahrenheit at $x$ hours after midnight. At what time of day is the temperature a maximum?

53. What is the temperature at 9 am? Round your answer to the nearest degree Fahrenheit.

54. The number of people who live in a seaside town can be modeled by the function $h(x) = 15{,}496 + 7500 \sin \frac{\pi}{26} (x - 12)$, where $h(x)$ is the number of people living in the town during the xth week of the year. What is the minimum population of the town?

_____

55. During what week of the year does the largest number of people live in the town?

_____

56. The average number of thunderstorms that occur in the xth month of the year in the town of Plaxon is modeled by the function $N(x) = 5 + 4 \sin \frac{\pi}{6} (x - 3)$. What is the average number of thunderstorms that occur in November in Plaxon? Round your answer to the nearest hundredth.

_____

57. During what month is the average number of thunderstorms occurring in Plaxon a maximum?

_____

# Answers to Chapter Questions

1. Answer: c  Objective: 1A

2. Answer: a  Objective: 1a

3. Answer: a  Objective: 2a

4. Answer: b  Objective: 4a

5. Answer: c  Objective: 4a

6. Answer: a  Objective: 5a

7. Answer: c. $\cos 60°$  Objective: 1a

8. Answer:

   b. $5\sqrt{2}$

   Objective: 1b

9. Answer: d. 8  Objective: 1b

10. Answer: b. $60°$  Objective: 1b

11. Answer:

    a. $2\sqrt{2}$

    Objective: 1b

12. Answer:

    a. $\left[\dfrac{1}{2}, \dfrac{\sqrt{3}}{2}\right]$

    Objective: 2a

13. Answer:

    c. $\left[-\dfrac{1}{\sqrt{2}}, -\dfrac{1}{\sqrt{2}}\right]$

    Objective: 2a

14. Answer: d. $(0.80, -0.60)$  Objective: 2a

15. Answer: d. $-315°$ and $45°$  Objective: 2b

16. Answer: c. $22.62°$  Objective: 3a

17. Answer: a. 17.01 m   Objective: 3a

18. Answer: d. 315°   Objective: 3b

19. Answer:
    c. $h(\theta) = 2 \sin \frac{3}{2}\theta$

    Objective: 4a

20. Answer:
    c. $h(\theta) = \sin \left[\frac{1}{2} \theta + 45\right]$

    Objective: 4a

21. Answer: b. vertical shift 1, phase shift –30   Objective: 4a

22. Answer: c. 143.99 ft   Objective: 6a

23. Answer: b. 47.89 in$^2$   Objective: 6a

24. Answer: c. .86   Objective: 6a

25. Answer: a. 179.05°   Objective: 6a

26. Answer: b. July   Objective: 7a

27. Answer: a. April and October   Objective: 7a

28. Answer:
    $x = \frac{17\sqrt{3}}{2}$

    Objective: 1b

29. Answer:
    $9\sqrt{3}$

    Objective: 1b

30. Answer:
    $\left[\frac{\sqrt{3}}{2}, -\frac{1}{2}\right]$

    Objective: 2a

31. Answer:
    $\left[-\frac{1}{\sqrt{2}}, \frac{1}{\sqrt{2}}\right]$
    Objective: 2a

32. Answer: (-0.62, 0.79)   Objective: 2a

33. Answer: Answers will vary. Examples: 40°, 400°, -320°, ...   Objective: 2b

34. Answer: 73.30°   Objective: 3a

35. Answer: 97.48 ft   Objective: 3a

36. Answer: 9   Objective: 3a

37. Answer: 120°   Objective: 3b

38. Answer: $g(\theta) = -\sin 2\theta$   Objective: 4a

39. Answer: 120°   Objective: 4a

40. Answer:
$$\text{amp} = \frac{1}{2} \; ; \; \text{period} = 90°$$

Objective: 4a

41. Answer:
$$\frac{\sqrt{3}}{2}$$

Objective: 5a

42. Answer:
$$\frac{1}{\sqrt{2}} = \frac{\sqrt{2}}{2}$$

Objective: 5a

43. Answer:
$$\frac{3\pi}{4}$$

Objective: 5a

44. Answer: $f(x) = 2 + \cos 4x$   Objective: 5b

45. Answer: $6\pi$   Objective: 5b

46. Answer: amplitude = 3; period = $3\pi$

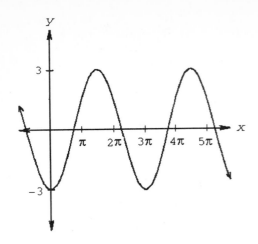

Objective: 5b

47. Answer:
    $-\dfrac{\pi}{3}$, 5.3
    Objective: 5b

48. Answer: 27.23 m  Objective: 6a

49. Answer: 220 cm$^2$  Objective: 6a

50. Answer: 18.75 ft; 140.625 ft$^2$  Objective: 6a

51. Answer: 12.57 in.; 100.53 in.$^2$  Objective: 6a

52. Answer: noon (12:00 pm)  Objective: 7a

53. Answer: 75°F  Objective: 7a

54. Answer: 7996  Objective: 7a

55. Answer: the 25th  Objective: 7a

56. Answer: 1.54  Objective: 7a

57. Answer: June  Objective: 7a

# CHAPTER 9

## QUANTITATIVE COMPARISON

In the space provided, write:
a. if the quantity of Column A is greater than the quantity in Column B;
b. if the quantity in Column B is greater than the quantity in Column A;
c. if the two quantities are equal; or
d. if the relationship cannot be determined from the information given.

| Column A | Column B | Answer |
|---|---|---|

1.

| $x$ when $y$ is 81, if $y$ varies inversely as the square of $x$, and $y$ is 1 when $x$ is 27 | $y$ when $x$ is 9, if $y$ varies inversely as the square of $x$, and $y$ is 1 when $x$ is 27 | _____ |
|---|---|---|

2.

| $\dfrac{1}{x^2 + 2}$ | $\dfrac{1}{x^2 + 3}$ | _____ |
|---|---|---|

3.

| $\dfrac{x}{x^2 + 2}$ | $\dfrac{x}{x^2 + 3}$ | _____ |
|---|---|---|

## MULTIPLE CHOICE   Circle the letter of the best answer choice.

4. If $y$ varies inversely as $x$, and $y$ is 6 when $x$ is 15, what is the equation of variation?

   a. $y = \dfrac{2}{5}x$      b. $y = x - 90$      c. $y = \dfrac{90}{x}$      d. $y = 21 - x$

5. If $y$ varies inversely as the square root of $x$, and $y$ is 5 when $x$ is 400, what is $x$ when $y$ is 2?

   a. 2500      b. $5\sqrt{2}$           c. 900           d. $500\sqrt{2}$

6. According to Newton's law of gravitation, the weight of an object varies inversely as the square of the distance from the object to the center of the Earth. The radius of the Earth is approximately 4000 mi. An astronaut weighs 150 lbs at the Earth's surface. How much does the same astronaut weigh at a height of 40 mi above the Earth's surface? Round your answer to the nearest pound.
   a. 123 lbs      b. 168 lbs        c. 139 lbs        d. 147 lbs

7. The intensity of light, $I$, measured in lux, from a light bulb varies inversely as the square of the distance, $d$, from the light bulb. If the intensity is 160 lux at a distance of 3 m, at what distance is the intensity 100 lux?
   a. 5.6 m        b. 4.8 m          c. 3.8 m          d. 2.4 m

8. What function is shown in the graph?

a. $f(x) = \dfrac{1}{x} + 1$

b. $g(x) = \dfrac{1}{x + 1}$

c. $h(x) = \dfrac{1}{x - 1}$

d. $k(x) = \dfrac{1}{x} - 1$

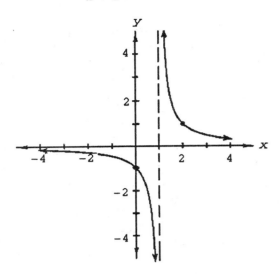

9. What function is shown in the graph?

a. $f(x) = \dfrac{1}{x + 1} - 3$

b. $g(x) = \dfrac{1}{x + 1} + 3$

c. $h(x) = \dfrac{1}{x - 1} - 3$

d. $k(x) = \dfrac{1}{x - 1} + 3$

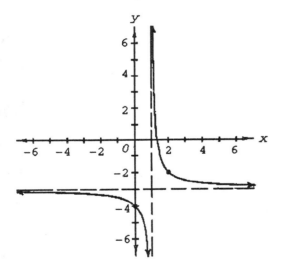

10. What are the asymptotes of $g(x) = \dfrac{1}{x - 3} - 4$?

a. $x = 3, \ y = 4$

b. $x = -3, \ y = -4$

c. $x = 3, \ y = -4$

d. $x = -3, \ y = 4$

11. Which graph shows $f(x) = \dfrac{3x + 10}{x + 3}$?

a.

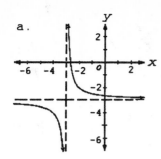

b.

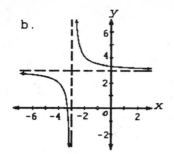

c.

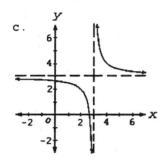

d.

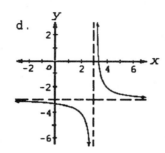

12. Which of the following functions is even?

a. $f(x) = \dfrac{1}{x - x^2}$        b. $g(x) = 1 + \dfrac{1}{x^2 - 5}$

c. $h(x) = \dfrac{x^2}{(x - 5)^2}$        d. all of these

13. Which of the following functions is odd?
I. $f(x) = x^3 + 3x^3 + 1$    II. $g(x) = 4x^3 - 1$    III. $h(x) = 5x^5 + x^3 - 2x$
a. I. only      b. I. and III.      c. I. and II.      d. III. only

14. Which of the following functions are even?

I. $f(x) = (x - 9)^2 + 4$        II. $g(x) = \dfrac{x^2 + 14}{5 - x^4}$

III. $h(x) = \dfrac{x - x^3}{x^2 + 5}$        IV. $k(x) = 11 + x^2$

a. I, II, III    b. II and IV    c. II, III, IV    d. I, II, IV

15. What are the vertical asymptotes of the function $g(x) = \dfrac{49}{x^2 - 7x - 30}$?

a. $x = -3,\ x = 10$        b. $x = 3,\ x = -10$
c. $x = -3,\ x = -10$        d. $x = 3,\ x = 10$

16. What function is sketched in the graph?

a. $f(x) = \dfrac{1}{(x+4)(x-2)}$

b. $g(x) = \dfrac{1}{(x+2)(x+4)}$

c. $h(x) = \dfrac{1}{(x-4)(x+2)}$

d. $k(x) = \dfrac{1}{(x-2)(x-4)}$

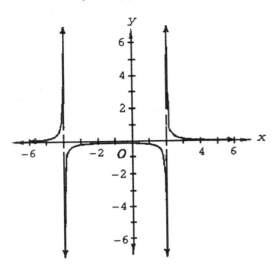

17. Which graph shows a sketch of the reciprocal of $f(x) = x^2 + 1$?

a.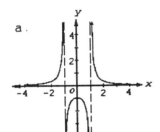

b.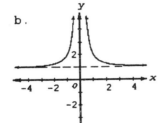

c.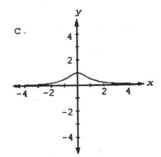

d.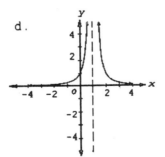

18. What is the horizontal asymptote of $f(x) = \dfrac{5x^2 - 3x + 12}{x^2 + 7x + 6}$?

a. $y = 5$  b. $y = 6$  c. $y = -1$  d. $y = -6$

19. Does the function $f(x) = \dfrac{3x^2 + 2x - 8}{x^2 - 3x - 10}$ have a hole in its graph? If so, where does the hole occur?

a. The graph of $f$ has no hole.

b. The graph of $f$ has a hole at $x = 2$.

c. The graph of $f$ has a hole at $x = -3$.

d. The graph of $f$ has a hole at $x = -2$.

20. Simplify the rational expression.

$$\frac{x^2}{2y} + \frac{2y}{x^2}$$

   a. $\dfrac{x^2 + 4y}{2y}$     b. $\dfrac{x^2 + 2y}{2yx^2}$     c. 1        d. $\dfrac{x^4 + 4y^2}{2x^2y}$

21. Simplify the rational expression.

$$\frac{x + 3}{x - 2} + \frac{x + 2}{x - 3}$$

   a. $\dfrac{2x^2 - 13}{x^2 - 5x + 6}$    b. $\dfrac{2x + 5}{2x - 5}$    c. $\dfrac{x^2 - 9}{x^2 + 4x + 4}$    d. $\dfrac{2x^2 + 12}{x^2 - 5x + 6}$

22. Simplify the rational expression.

$$\frac{2x}{x - 3} - \frac{x}{x + 2}$$

   a. $\dfrac{x^2 - x}{x^2 + 5x - 6}$   b. $\dfrac{x^2 - 4x}{x^2 - 5x + 6}$   c. $\dfrac{x^2 + x}{x^2 - x + 6}$   d. $\dfrac{x^2 + 7x}{x^2 - x - 6}$

23. Solve the rational equation.

$$\frac{1}{x + 1} + \frac{2}{x + 2} = \frac{3}{x + 3}$$

   a. $x = -\dfrac{2}{3}$     b. $x = \dfrac{7}{8}$     c. $x = -\dfrac{3}{2}$     d. $x = \dfrac{3}{5}$

***SHORT ANSWER***   ***Write the answer in the space provided.***

24. If $y$ varies inversely as the square of $x$, and $y$ is 1 when $x$ is 12, what is $y$ when $x$ is 6?

     _____

25. If $y$ varies inversely as the cube of $x$, and $y$ is 7 when $x$ is 3, what is the equation of variation?

     _____

26. If the area of a triangle is held constant, the length of the base varies inversely as the height. Write the inverse relation equation for a triangular area of 57 square in.

     _____

27. If $y$ varies inversely as $x$, and $y$ is 5000 when $x$ is $\frac{1}{2}$, what is $x$ when $y$ is 50?

     _____

28. According to Boyle's law, the volume, $V$, of a gas at constant temperature varies inversely as the pressure, $P$, of the gas. If the gas volume is 420 cubic ft when the gas pressure is 20 lbs per square in., write the equation of variation.

     _____

29. What function is shown in the graph?

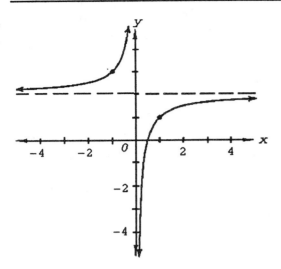

30. What function is shown in the graph?

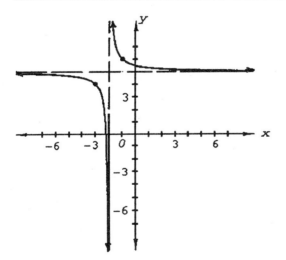

31. Use polynomial division to write $f(x) = \dfrac{6x - 23}{x - 4}$ in the form $f(x) = \dfrac{1}{x - a} + b$.

32. What are the asymptotes of $h(x) = \dfrac{1}{x + 5} + 2$?

33. Use polynomial division to write $f(x) = \dfrac{2x + 7}{x + 3}$ in the form $f(x) = \dfrac{1}{x - a} + b$.

34. What are the asymptotes of $f(x) = \dfrac{2x + 7}{x + 3}$?

_____

35. Sketch the graph of $f(x) = \dfrac{2x + 7}{x + 3}$. Show the asymptotes. Label at least 3 points on the graph.

36. Is the function $g(x) = 6 + \dfrac{1}{x}$ even, odd, or neither?

_____

37. Let $f(x)$ be an even function. If $f(6) = \dfrac{15}{2}$, what is $f(-6)$?

_____

38. Sketch the graph of the reciprocal of $f(x) = x^2 - 4$. Show the vertical asymptotes and label 3 points on the graph.

39. What are the vertical asymptotes of $h(x) = \dfrac{-5}{x(x - 3)}$?

_____

40. What are the vertical asymptotes of $f(x) = \dfrac{3}{x^2 - 2x - 15}$?

_____

41. Let $f(x)$ be the function sketched in the graph. What are the vertical asymptotes of the graph of the reciprocal of $f$?

_____

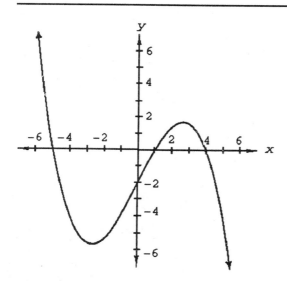

42. Simplify the rational function.
$$f(x) = \dfrac{x^2 - 3x - 28}{x^2 - 49}$$

_____

43. Find the vertical and horizontal asymptotes of $f(x) = \dfrac{x}{4x^2 - 100}$.

_____

44. What is the domain of the function $g(x) = \dfrac{x^2 - 7x}{x^2 - 49}$?

_____

45. Find the asymptotes of the function $h(x) = \dfrac{4x^3}{x^3 - 5x}$.

_____

46. Find the vertical and horizontal asymptotes of $f(x) = \dfrac{26x + 9}{x^2 + 11x + 30}$.

_____

47. Describe what causes a hole in a graph.

_____

48. Simplify the rational expression.

$\dfrac{1}{x} + \dfrac{1}{2x} + \dfrac{3}{x^2}$

_____

49. Simplify the rational expression.

$\dfrac{3}{x - 2} + \dfrac{6}{x + 2}$

_____

50. Solve the rational equation.

$\dfrac{4x^2 + 2x}{x^2 - 5x} = 2$

_____

51. Solve the rational equation.

$\dfrac{-4}{9x + 8} = \dfrac{2}{3x - 9}$

_____

52. Find all the solutions of the rational equation.

$\dfrac{1}{x} + \dfrac{1}{x^2} = \dfrac{2}{x^3}$

_____

53. Solve the rational equation.

$\dfrac{-4}{x^2 + 2x} = \dfrac{1}{2x + 4}$

_____

# *Answers to Chapter Questions*

1. Answer: b   Objective: 1A

2. Answer: a   Objective: 5b

3. Answer: d   Objective: 5b

4. Answer:

   c. $y = \dfrac{90}{x}$

  Objective: 1a

5. Answer: a. 2500   Objective: 1a

6. Answer: d. 147 lbs   Objective: 1a

7. Answer: c. 3.8 m   Objective: 1a

8. Answer:

   c. $h(x) = \dfrac{1}{x - 1}$

  Objective: 2a

9. Answer:

   c. $h(x) = \dfrac{1}{x - 1} - 3$

  Objective: 2a

10. Answer: c. $x = 3,\ y = -4$   Objective: 2a

11. Answer:

   b.

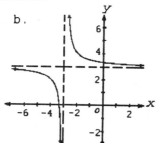

  Objective: 2a

12. Answer:

   b. $g(x) = 1 + \dfrac{1}{x^2 - 5}$

  Objective: 3a

13. Answer: d. III only   Objective: 3a

14. Answer: b. II. and IV.   Objective: 3a

15. Answer: a. $x = -3$, $x = 10$   Objective: 3b

16. Answer:

a. $f(x) = \dfrac{1}{(x + 4)(x - 2)}$

Objective: 3b

17. Answer:

c.

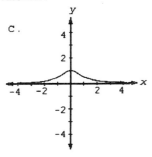

Objective: 3b

18. Answer: a. $y = 5$   Objective: 4a

19. Answer: d. The graph of $f$ has a hole at $x = -2$.   Objective: 4a

20. Answer:

$\dfrac{x^4 + 4y^2}{2x^2 y}$

Objective: 5a

21. Answer:

d. $\dfrac{2x^2 - 13}{x^2 - 5x + 6}$

Objective: 5a

22. Answer:

d. $\dfrac{x^2 + 7x}{x^2 - x - 6}$

Objective: 5a

23. Answer:

c. $x = -\dfrac{3}{2}$

Objective: 5b

24. Answer: 4  Objective: 1a

25. Answer:
$$y = \frac{189}{x^3}$$
Objective: 1a

26. Answer:
$$b = \frac{114}{h}, \text{ or any equivalent equation}$$
Objective: 1a

27. Answer: 50  Objective: 1a

28. Answer:
$$V = \frac{8400}{P}, \text{ or any equivalent equation}$$
Objective: 1a

29. Answer:
$$f(x) = 2 - \frac{1}{x}$$
Objective: 2a

30. Answer:
$$f(x) = \frac{1}{x + 2} + 5$$
Objective: 2a

31. Answer:
$$f(x) = \frac{1}{x - 4} + 6$$
Objective: 2a

32. Answer: $x = -5$, $y = 2$  Objective: 2a

33. Answer:
$$f(x) = \frac{1}{x + 3} + 2$$
Objective: 2a

34. Answer: $x = -3$, $y = 2$  Objective: 2a

35. Answer:

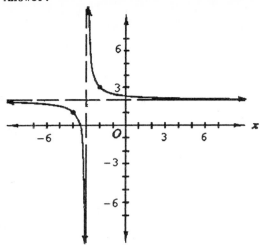

Objective: 2a

36. Answer: neither  Objective: 3a

37. Answer:
$$\frac{15}{2}$$

Objective: 3a

38. Answer:

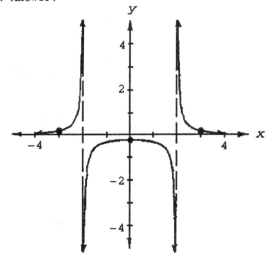

Objective: 3b

39. Answer: $x = 0$, $x = 3$  Objective: 3b

40. Answer: $x = -3$, $x = 5$  Objective: 3b

41. Answer: $x = 4$, $x = 1$, $x = -5$  Objective: 3b

42. Answer:
$$f(x) = \frac{x + 4}{x + 7}, \text{ for } x \neq 7 \text{ and } x \neq -7$$
Objective: 4a

43. Answer: $x = 5$, $x = -5$, $y = 0$   Objective: 4a

44. Answer: all real numbers except 7 and -7   Objective: 4a

45. Answer: $x = 5$, $x = 0$, $y = 4$   Objective: 4a

46. Answer: $y = 0$, $x = -5$, $x = -6$   Objective: 4a

47. Answer:
Answers will vary. Ex.: A hole would appear at $x = 5$ if $(x - 5)$ is a factor of the numerator and the denominator of a function.
Objective: 4a

48. Answer:
$$\frac{3x + 6}{2x^2}$$
Objective: 5a

49. Answer:
$$\frac{9x - 6}{x^2 - 4}$$
Objective: 5a

50. Answer: $x = -6$   Objective: 5b

51. Answer:
$$x = \frac{2}{3}$$
Objective: 5b

52. Answer: $x = 1$ or $x = -2$   Objective: 5b

53. Answer: $x = -8$   Objective: 5b

# CHAPTER 10

## QUANTITATIVE COMPARISON

In the space provided, write:
a. if the quantity of Column A is greater than the quantity in Column B;
b. if the quantity in Column B is greater than the quantity in Column A;
c. if the two quantities are equal; or
d. if the relationship cannot be determined from the information given.

| Column A | Column B | Answer |
|---|---|---|

1.

| x-coordinate of focus of parabola with equation $(y - 2)^2 = 14\left(x + \dfrac{1}{2}\right)$ | y-coordinate of focus of parabola with equation $(y - 2)^2 = 14\left(x + \dfrac{1}{2}\right)$ | _____ |
|---|---|---|

2.

| x-coordinate of the center of the circle represented by the equation $(x + 1)^2 + y^2 = 1$ | y-coordinate of the center of the circle represented by the equation $(x + 1)^2 + y^2 = 1$ | _____ |
|---|---|---|

3.

| x-coordinate of the center of the hyperbola $\dfrac{(x - 2)^2}{9} - \dfrac{(y + 5)^2}{1} = 1$ | y-coordinate of the center of the hyperbola $\dfrac{(x - 2)^2}{9} - \dfrac{(y + 5)^2}{1} = 1$ | _____ |
|---|---|---|

## MULTIPLE CHOICE   *Circle the letter of the best answer choice.*

4. Determine an equation of the parabola with focus $\left[0, \dfrac{1}{2}\right]$ and directrix $x = -\dfrac{1}{2}$.

    a. $x = \dfrac{1}{2}y^2$     b. $x = -\dfrac{1}{4}y^2$       c. $x = \dfrac{1}{8}y^2$         d. $x = -\dfrac{1}{2}y^2$

5. Determine the vertex, focus, and directrix for the parabola $y = x^2$.

    a. vertex $(0, 0)$; focus $\left[0, \dfrac{1}{4}\right]$; directrix $y = -\dfrac{1}{4}$

    b. vertex $(1, 0)$; focus $\left[0, \dfrac{1}{2}\right]$; directrix $y = -\dfrac{1}{2}$

    c. vertex $(0, 1)$; focus $\left[0, -\dfrac{1}{2}\right]$; directrix $y = \dfrac{1}{2}$

    d. vertex $(0, 0)$; focus $\left[0, -\dfrac{1}{4}\right]$; directrix $y = \dfrac{1}{4}$

6. Determine the vertex, focus, and directrix for the parabola $x = \frac{1}{16}y^2$.

    a. vertex $(0, -1)$; focus $(-4, 0)$; directrix $x = -2$
    b. vertex $(-1, 0)$; focus $(4, 0)$; directrix $x = 3$
    c. vertex $(0, 0)$; focus $(-4, 0)$; directrix $x = 4$
    d. vertex $(0, 0)$; focus $(4, 0)$; directrix $x = -4$

7. Write the equation, in standard form, of the circle with center $C(-3, -4)$ and radius $r = 7$.
    a. $(x - 3)^2 + (y + 4)^2 = 7$      b. $(x + 3)^2 + (y - 4)^2 = 49$
    c. $(x + 3)^2 + (y + 4)^2 = 49$      d. $(x - 3)^2 + (x - 4)^2 = 49$

8. Write the equation, in standard form, of the circle with center $C(-3, 5)$ and radius $r = 1$.
    a. $(x + 3)^2 + (y - 5)^2 = 1$      b. $(x - 3)^2 + (y + 5)^2 = 1$
    c. $(x - 3)^2 + (y - 5)^2 = 1$      d. $(x + 3)^2 + (y + 5)^2 = 1$

9. Write the equation, in standard form, of the circle with center $C(a, 1)$ and radius $r = 3$.
    a. $(x + a)^2 + (y + 1)^2 = 9$      b. $x^2 + (y + 1)^2 = 9$
    c. $x^2 + (y - 1)^2 = 9$      d. $(x - a)^2 + (y - 1)^2 = 9$

10. Determine the center and radius of the circle represented by the equation $x^2 + (y + 4)^2 = \frac{3}{25}$.

    a. $C(0, -4)$, $r = \frac{\sqrt{3}}{5}$      b. $C(0, 4)$, $r = \frac{3}{5}$

    c. $C(4, 0)$, $r = \frac{5}{3}$      d. $C(-4, 0)$, $r = -\frac{3}{5}$

11. Determine the center and radius of the circle represented by the equation $(x + 3)^2 + (y + 5)^2 = 25$.
    a. $C(3, -5)$, $r = 25$      b. $C(-3, 5)$, $r = 10$
    c. $C(3, 5)$, $r = 5$      d. $C(-3, -5)$, $r = 5$

12. Determine the center and radius of the circle represented by the equation $2x^2 + 2y^2 + 4x + 8y = 8$.
    a. $C(1, 2)$, $r = 3$      b. $C(-1, -2)$, $r = 3$
    c. $C(-1, 2)$, $r = 3$      d. $C(1, -2)$, $r = 3$

13. Find the vertices and the foci for the ellipse $2x^2 + 3y^2 = 6$.
    a. vertices $[-\sqrt{2}, 0]$, $[\sqrt{2}, 0]$, $[0, -\sqrt{3}]$, $[0, \sqrt{3}]$; foci $(1, 0)$, $(0, -1)$

    b. vertices $[-\sqrt{3}, 0]$, $[\sqrt{3}, 0]$, $[0, -\sqrt{2}]$, $[0, \sqrt{2}]$; foci $(-1, 0)$, $(1, 0)$

    c. vertices $[-\sqrt{5}, 0]$, $[\sqrt{5}, 0]$, $[0, -\sqrt{7}]$, $[0, \sqrt{7}]$; foci $(2, 0)$, $(-2, 0)$

    d. vertices $(1, 0)$, $(-1, 0)$, $[0, -\sqrt{2}]$, $[0, \sqrt{2}]$ ; foci $(3, 0)$, $(-3, 0)$

14. Find the vertices and the foci for the ellipse $4x^2 + 9y^2 = 1$.

   a. vertices $(1, 0)$, $(-1, 0)$, $(0, -3)$, $(0, 3)$; foci $\left[-\dfrac{\sqrt{2}}{3}, 0\right]$, $\left[\dfrac{\sqrt{2}}{3}, 0\right]$

   b. vertices $(-2, 0)$, $(2, 0)$, $(0, -6)$, $(0, 6)$; foci $\left[\dfrac{\sqrt{3}}{5}, 0\right]$, $\left[-\dfrac{\sqrt{3}}{5}, 0\right]$

   c. vertices $\left[-\dfrac{1}{2}, 0\right]$, $\left[\dfrac{1}{2}, 0\right]$, $\left[0, -\dfrac{1}{3}\right]$, $\left[0, \dfrac{1}{3}\right]$; foci $\left[-\dfrac{\sqrt{5}}{6}, 0\right]$, $\left[\dfrac{\sqrt{5}}{6}, 0\right]$

   d. vertices $\left[-\dfrac{3}{4}, 0\right]$, $\left[\dfrac{3}{4}, 0\right]$, $\left[0, -\dfrac{2}{3}\right]$, $\left[0, \dfrac{2}{3}\right]$; foci $\left[\dfrac{3}{4}, 0\right]$, $\left[-\dfrac{3}{4}, 0\right]$

15. Find the center of the ellipse $4x^2 + y^2 - 8x - 2y + 1 = 0$.
   a. $C(1, 1)$        b. $C(0, 0)$        c. $C(2, 2)$        d. $C(-1, -1)$

16. Find the center and vertices of the hyperbola $\dfrac{(y + 3)^2}{4} - \dfrac{(x + 1)^2}{16} = 1$.

   a. $C(-1, -3)$; vertices $(-1, -5)$, $(-1, -1)$
   b. $C(1, 3)$; vertices $(1, 5)$, $(1, 1)$
   c. $C(1, -3)$; vertices $(1, -5)$, $(1, -1)$
   d. $C(-1, 3)$; vertices $(-1, 5)$, $(-1, 1)$

17. Find the center, the vertices, and the foci of the hyperbola $x^2 - 4y^2 = 4$.

   a. $C(0, 1)$; vertices $(0, -2)$, $(0, 2)$; foci $\left[0, \sqrt{5}\right]$, $\left[0, -\sqrt{5}\right]$

   b. $C(0, 0)$; vertices $(0, -2)$, $(-2, 0)$; foci $\left[-\sqrt{3}, 0\right]$, $\left[\sqrt{3}, 0\right]$

   c. $C(0, -1)$; vertices $(2, 0)$, $(0, -2)$; foci $\left[0, \sqrt{3}\right]$, $\left[0, -\sqrt{3}\right]$

   d. $C(0, 0)$; vertices $(-2, 0)$, $(2, 0)$; foci $\left[-\sqrt{5}, 0\right]$, $\left[\sqrt{5}, 0\right]$

18. Choose the correct graph for the hyperbola $\dfrac{x^2}{4} - \dfrac{y^2}{9} = 1$.

a.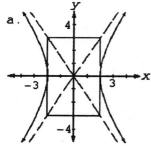

b.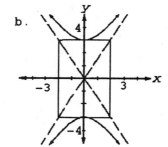

c.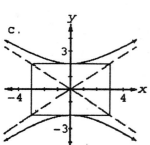

d.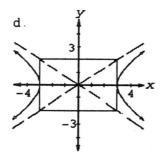

19. Solve the system: $\begin{cases} y^2 - 5y - 4x - 28 = 0 \\ y - 4x = 1 \end{cases}$
   a. $(7, -2)$ and $(6, -5)$        b. $(2, 9)$ and $(-1, -3)$
   c. $(4, 3)$ and $(8, 3)$        d. $(1, -2)$ and $(-5, 6)$

20. Solve the system: $\begin{cases} y = x^2 - 2x - 1 \\ y = x + 3 \end{cases}$

    a. (5, 7) and (1, -2)            b. (3, 2) and (3, -7)

    c. (3, 5) and (4, 8)            d. (4, 7) and (-1, 2)

21. Solve the system: $\begin{cases} 2x^2 + 5y^2 = 22 \\ 3x^2 - y^2 = -1 \end{cases}$

    a. (1, 2), (1, -2), (-1, 2), (-1, -2)

    b. (5, 6), (5, -6), (-5, 6), (-5, -6)

    c. (3, 4), (3, -4), (-3, 4), (-3, -4)

    d. $\left[\frac{1}{2}, 1\right]$, $\left[\frac{1}{2}, -1\right]$, $\left[-\frac{1}{2}, 1\right]$, $\left[-\frac{1}{2}, -1\right]$

22. Write a parametric representation and name the geometric object represented by $(x - 1)^2 + (y - 1)^2 = 1$.

    a. $x = 3 \cos t$, $y = -3 \sin t$; a circle

    b. $x = 1 + \cos t$, $y = 1 + \sin t$; a circle

    c. $x = -\cos t$, $y = \sin^2 t$; a circle

    d. $x = \sin t$, $y = -\sin t$; a circle

23. Which parametric representation and name of the geometric object represents $x + y = 7$?

    a. $x = 3 + t$, $y = 2 - t$; a circle

    b. $x = 4 + t$, $y = 1 - t$; an ellipse

    c. $x = 5 + t$, $y = 3 - t$; a hyperbola

    d. $x = 5 - t$, $y = 2 + t$; a line

24. Which parametric representation and name of the geometric object represents $x + y = 0$ where $-1 \le x \le 1$?

    a. $x = \sin t$, $y = \cos t$; a hyperbola

    b. $x = \cos t$, $y = \sin t$; an ellipse

    c. $x = \sin t$, $y = -\sin t$; a line

    d. $x = \cos t$, $y = -\cos t$; a circle

**SHORT ANSWER**    *Write the answer in the space provided.*

25. Determine an equation of the parabola with focus (3, 0) and directrix $x = -3$.

_____

26. Determine an equation of the parabola with focus (-6, 0) and directrix $x = 6$.

_____

27. Determine an equation of the parabola with focus (0, 4) and directrix $y = -4$.

_____

28. Determine the vertex, focus, and directrix for the following parabola and sketch the graph: $y = \frac{1}{8}x^2$.

_____

29. Determine the vertex, focus, and directrix for the following parabola and sketch the graph: $x = \frac{1}{12}y^2$.

_____

30. Determine the vertex, focus, and directrix from the graph.

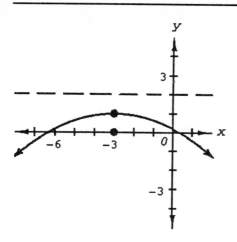

31. Write the equation, in standard form, of the circle with center $C(0, 1)$ and radius $r = 2$.

32. Write the equation, in standard form, of the circle with center $C(2, -2)$ and radius $r = \dfrac{1}{2}$.

33. Determine the center and radius of the circle represented by the equation $x^2 + y^2 = 16$.

34. Find the vertices and the foci for the ellipse $\dfrac{x^2}{4} + \dfrac{y^2}{1} = 1$.

35. Find the vertices and the foci for the ellipse $16x^2 + 9y^2 = 144$.

36. Find the equation of an ellipse given vertices $(-2, 0)$, $(2, 0)$, and co-vertices, $\left[0, -\sqrt{2}\right]$, $\left[0, \sqrt{2}\right]$.

37. Write the equation of an ellipse given vertices $(0, 6)$, $(0, -6)$, and co-vertices $(4, 0)$, $(-4, 0)$.

38. Find the center, vertices and the foci of the ellipse $4x^2 + 9y^2 - 16x + 18y - 11 = 0$.

39. Determine the center, the vertices, and the foci of the hyperbola. Then graph the hyperbola.

$$\frac{x^2}{9} - \frac{y^2}{1} = 1$$

---

40. Find the center, the vertices, and the foci of the hyperbola $4y^2 - x^2 = 4$.

---

41. Find the vertices and the foci and graph the hyperbola $25y^2 - 16x^2 = 400$.

---

42. Find the center, the vertices, and the foci and graph the hyperbola $4x^2 - y^2 + 24x + 4y + 28 = 0$.

---

43. Given the center $(1, -2)$, vertices $(-4, -2)$, $(6, -2)$, and foci $\left[1 - \sqrt{29}, -2\right]$, $\left[1 + \sqrt{29}, -2\right]$, write the equation of the hyperbola.

---

44. Find the center, the vertices, and the foci of the hyperbola $x^2 - y^2 - 2x - 4y - 4 = 0$.

---

45. Solve the system graphically: $\begin{cases} x^2 + y^2 = 25 \\ 3x - 4y = 0 \end{cases}$

---

46. Solve the system graphically: $\begin{cases} y = 2x + 6 \\ x^2 = 2y \end{cases}$

---

47. Solve the system graphically: $\begin{cases} x^2 + y^2 = 25 \\ \dfrac{x^2}{25} - \dfrac{y^2}{25} = 1 \end{cases}$

---

48. Solve the system graphically: $\begin{cases} 9x^2 + 16y^2 = 144 \\ 3x^2 + 4y^2 = 36 \end{cases}$

---

49. Solve the system graphically: $\begin{cases} x^2 + y^2 = 25 \\ (3x - 4y)(3x + 4y) = 0 \end{cases}$

---

50. Write a parametric representation and name the geometric object represented by $x^2 + y^2 = 4$.

---

51. Write a parametric representation and name the geometric object represented by $x + y = 3$ where $0 \leq x \leq 2$.

_____

52. Write a parametric representation and name the geometric object represented by $x^2 + \dfrac{y^2}{4} = 1$.

_____

53. Write a rectangular representation for the following system. Find the center.
$$\begin{cases} x(t) = 6 \cos t \\ y(t) = 3 \sin t \end{cases}$$

_____

54. Write a parametric representation and name the geometric object represented by $x^2 + y^2 = 1$, $x \geq 0$ and $y \geq 0$.

_____

# Answers to Chapter Questions

1. Answer: a   Objective: 1A

2. Answer: b   Objective: 2b

3. Answer: a   Objective: 4a

4. Answer:
   a. $x = \frac{1}{2}y^2$

   Objective: 1a

5. Answer:
   a. vertex $(0, 0)$; focus $\left[0, \frac{1}{4}\right]$; directrix $y = -\frac{1}{4}$

   Objective: 1b

6. Answer: d. vertex $(0, 0)$; focus $(4, 0)$; directrix $x = -4$   Objective: 1b

7. Answer: c. $(x + 3)^2 + (y + 4)^2 = 49$   Objective: 2a

8. Answer: a. $(x + 3)^2 + (y - 5)^2 = 1$   Objective: 2a

9. Answer: d. $(x - a)^2 + (y - 1)^2 = 9$   Objective: 2a

10. Answer:
    a. $C(0, -4)$, $r = \frac{\sqrt{3}}{5}$

    Objective: 2b

11. Answer: d. $C(-3, -5)$, $r = 5$   Objective: 2b

12. Answer: b. $C(-1, -2)$, $r = 3$   Objective: 2b

13. Answer:
    b. vertices $\left[-\sqrt{3}, 0\right]$, $\left[\sqrt{3}, 0\right]$, $\left[0, -\sqrt{2}\right]$; foci $(-1, 0)$, $(1, 0)$

    Objective: 3a

14. Answer:
    c. vertices $\left[-\frac{1}{2}, 0\right]$, $\left[\frac{1}{2}, 0\right]$, $\left[0, -\frac{1}{3}\right]$, $\left[0, \frac{1}{3}\right]$; foci $\left[-\frac{\sqrt{5}}{6}, 0\right]$, $\left[\frac{\sqrt{5}}{6}, 0\right]$

    Objective: 3a

15. Answer: a. $C(1, 1)$  Objective: 3a

16. Answer: a. $C(-1, -3)$; vertices $(-1, -5)$, $(-1, -1)$  Objective: 4a

17. Answer:

   d. $C(0, 0)$; vertices $(-2, 0)$, $(2, 0)$; foci $\left[-\sqrt{5},\ 0\right]$, $\left[\sqrt{5},\ 0\right]$

   Objective: 4a

18. Answer:

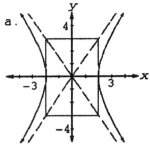

a.

Objective: 4a

19. Answer: b. $(2, 9)$ and $(-1, -3)$  Objective: 5a

20. Answer: d. $(4, 7)$ and $(-1, 2)$  Objective: 5a

21. Answer: a. $(1, 2)$, $(1, -2)$, $(-1, 2)$, $(-1, -2)$  Objective: 5b

22. Answer: b. $x = 1 + \cos t$, $y = 1 + \sin t$; a circle  Objective: 6a

23. Answer: d. $x = 5 - t$, $y = 2 + t$; a line  Objective: 6a

24. Answer: c. $x = \sin t$, $y = -\sin t$; a line  Objective: 6b

25. Answer:

   $x = \dfrac{1}{12}y^2$

   Objective: 1a

26. Answer:

   $x = -\dfrac{1}{24}y^2$

   Objective: 1a

27. Answer:

   $y = \dfrac{1}{16}x^2$

   Objective: 1a

28. Answer: vertex $(0, 0)$; focus $(0, 2)$; directrix $y = -2$

144

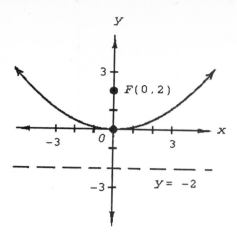

Objective: 1b

29. Answer: vertex (0, 0); focus (-3, 0); directrix $x = 3$

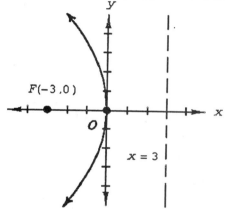

Objective: 1b

30. Answer: vertex (-3, 1); focus (-3, 0); directrix $y = 2$   Objective: 1b

31. Answer: $x^2 + (y - 1)^2 = 4$   Objective: 2a

32. Answer:

$$(x - 2)^2 + (y + 2)^2 = \frac{1}{4}$$

Objective: 2a

33. Answer: $C(0, 0)$, $r = 4$   Objective: 2b

34. Answer:

vertices (-2, 0), (2, 0), (0, -1), (0, 1); foci $\left[-\sqrt{3}, 0\right]$, $\left[\sqrt{3}, 0\right]$

Objective: 3a

35. Answer:
   vertices $(-3, 0)$, $(3, 0)$, $(0, -4)$, $(0, 4)$; foci $\left[0, -\sqrt{7}\right]$, $\left[0, \sqrt{7}\right]$

   Objective: 3a

36. Answer:
   $x^2 + 2y^2 = 4$ or $\dfrac{x^2}{4} + \dfrac{y^2}{2} = 1$

   Objective: 3a

37. Answer:
   $\dfrac{x^2}{16} + \dfrac{y^2}{36} = 1$

   Objective: 3a

38. Answer:
   $C(2, -1)$; vertices $(-1, -1)$, $(5, -1)$, $(2, -3)$, $(2, 1)$; foci
   $\left[2 - \sqrt{5}, -1\right]$, $\left[2 + \sqrt{5}, -1\right]$

   Objective: 3a

39. Answer:
   $C(0, 0)$; vertices $(-3, 0)$, $(3, 0)$; foci $\left[-\sqrt{10}, 0\right]$, $\left[\sqrt{10}, 0\right]$;
   asymptotes $y = -\dfrac{1}{3}x$, $y = \dfrac{1}{3}x$

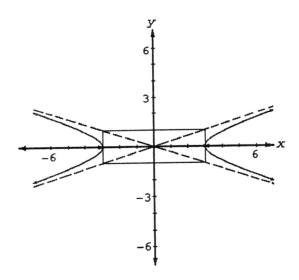

   Objective: 4a

40. Answer:
   $C(0, 0)$; vertices $(0, -1)$ $(0, 1)$; foci $\left[0, -\sqrt{5}\right]$, $\left[0, \sqrt{5}\right]$

   Objective: 4a

146

41. Answer:

vertices $(0, 4)$, $(0, -4)$; foci $\left[0, \sqrt{41}\right]$, $\left[0, -\sqrt{41}\right]$

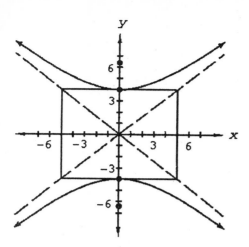

Objective: 4a

42. Answer:

$C(-3, 2)$; vertices $(-2, 2)$, $(-4, 2)$; foci $\left[-3 + \sqrt{5}, 2\right]$, $\left[-3 - \sqrt{5}, 2\right]$

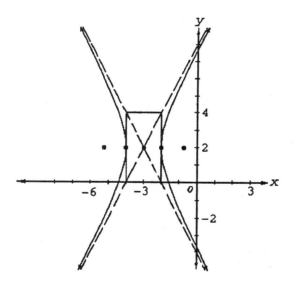

Objective: 4a

43. Answer:

$$\frac{(x - 1)^2}{25} - \frac{(y + 2)^2}{4} = 1$$

Objective: 4a

44. Answer:

 $C(1, -2)$; vertices $(0, -2)$, $(2, -2)$; foci $\left[1 - \sqrt{2}, -2\right]$, $\left[1 + \sqrt{2}, -2\right]$

 Objective: 4a


45. Answer: $(4, 3)$ and $(-4, -3)$

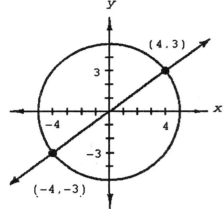

 Objective: 5a


46. Answer: $(-2, 2)$ and $(6, 18)$

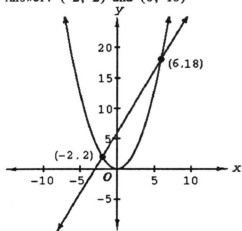

 Objective: 5a


47. Answer: $(5, 0)$ and $(-5, 0)$

148

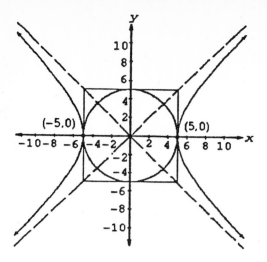

Objective: 5b

48. Answer: (0, 3) and (0, -3)

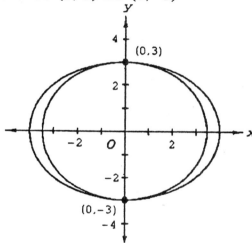

Objective: 5b

49. Answer: (4, 3), (4, -3), (-4, 3), (-4, -3)

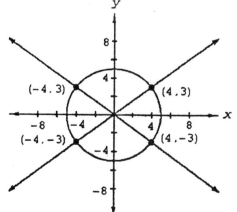

Objective: 5b

149

50. Answer:
    Answers for parametric representation may vary.
    $x = 2 \cos t$, $y = 2 \sin t$, a circle

    Objective: 6a


51. Answer:
    Answers for parametric representation may vary.
    $x = 1 + \cos t$, $y = 2 - \cos t$; a line segment

    Objective: 6a


52. Answer:
    Answers for parametric representation may vary.
    $x = \cos t$, $y = 2 \sin t$; an ellipse

    Objective: 6b


53. Answer:
    $\dfrac{x^2}{36} + \dfrac{y^2}{9}$, $(0, 0)$

    Objective: 6b


54. Answer:
    Answers for parametric representation may vary.
    $x = \sqrt{t}$, $y = \sqrt{1 - t}$; a quarter circle

    Objective: 6b

# CHAPTER 11

## QUANTITATIVE COMPARISON

In the space provided, write:
a. if the quantity of Column A is greater than the quantity in Column B;
b. if the quantity in Column B is greater than the quantity in Column A;
c. if the two quantities are equal; or
d. if the relationship cannot be determined from the information given.

| Column A | Column B | Answer |
|---|---|---|

1. 
| probability of rolling a 6 on a number cube | probability of rolling a 1 on a number cube | _____ |

2. 
| $P(E_1)$ given that $P(E_2|E_1) = 9$ and $P(E_1 \text{ and } E_2) = 6$ | $P(E_2)$ given that $P(E_2|E_1) = 9$ and $P(E_1 \text{ and } E_2) = 6$ | _____ |

3. 
| $_{12}C_4$ | $_{12}C_8$ | _____ |

4. 
| $_{15}C_4$ | $_{15}P_4$ | _____ |

## MULTIPLE CHOICE   Circle the letter of the best answer choice.

5. What is the probability of rolling an even number on a number cube?
   a. $\frac{1}{2}$     b. $\frac{2}{3}$     c. $\frac{4}{6}$     d. $\frac{5}{6}$

6. What is the probability of rolling a prime number on a number cube?
   a. $\frac{5}{6}$     b. $\frac{1}{6}$     c. $\frac{2}{3}$     d. $\frac{1}{2}$

7. What is the probability of drawing a black card from a well-shuffled standard deck of 52 cards?
   a. $\frac{17}{26}$     b. $\frac{1}{2}$     c. $\frac{2}{3}$     d. $\frac{4}{9}$

8. In how many ways can the letters in the word NUMBER be arranged if each letter is used exactly once in each arrangement?
   a. 540     b. 720     c. 360     d. 120

9. In how many ways can a president, a vice-president, a secretary, and a treasurer be chosen in a club containing 12 members?
   a. 9876     b. 10,345     c. 12,000     d. 11,880

10. A group of 3 men and 3 women wants to be photographed. In how many ways can they be arranged if they stand in 2 rows with the women in front?
   a. 12     b. 72     c. 36     d. 120

11. A real-estate development has 10 look-alike houses on 1 street. The contractor decides to paint 3 of them blue, 3 yellow, and 4 red. How many ways can this be done?
a. 3800  b. 4200  c. 3600  d. 4000

12. If all the letters are used, how many different permutations are possible using the letters of the word INFINITE?
a. 3360  b. 3280  c. 3490  d. 3120

13. What expression would you use to calculate how many different ways 12 swimmers could be arranged around a pool?
a. 10!  b. 11!  c. 12!  d. 13!

14. Evaluate $_5C_3$.
a. 10  b. 12  c. 7  d. 5

15. If 5 points lie on the circumference of a circle, how many inscribed triangles can be drawn having those points as vertices?
a. 100  b. 10  c. 25  d. 5

16. A bag contains 5 balls, 3 red and 2 green. A ball is drawn, replaced in the bag, and a ball is drawn a second time. What is the probability that both balls drawn are red?
a. $\frac{2}{5}$  b. $\frac{9}{25}$  c. $\frac{2}{10}$  d. $\frac{3}{5}$

17. Two number cubes are rolled. Find the probability that the 2 numbers thrown are equal or that they have total 8.
a. $\frac{5}{18}$  b. $\frac{2}{9}$  c. $\frac{4}{13}$  d. $\frac{7}{12}$

18. $P(E_2 \mid E_1) =$
a. $\frac{P(E_2)}{P(E_1)}$  b. $P(E_1 \text{ and } E_2)$  c. $\frac{P(E_1 \text{ and } E_2)}{P(E_1)}$  d. $\frac{P(E_1 \text{ and } E_2)}{P(E_2)}$

19. The notation $P(A \mid B)$ stands for
a. the conditional probability of event $A$ given event $B$.
b. the conditional probability of event $B$ given event $A$.
c. the conditional probability of event $A$ divided by the conditional probability of event $B$.
d. none of these

**SHORT ANSWER**  *Write the answer in the space provided.*

20. What is the probability of drawing an ace from a well-shuffled standard deck of 52 cards?

_____

21. Probabilities that are based on the results of polls or research are called _____.

_____

22. Karen tossed a coin 10 times and got heads 7 times. She would conclude that the experimental probability of heads is _____.

_____

| Color | Number of cars |
|-------|----------------|
| red | 7 |
| white | 9 |
| black | 3 |
| blue | 1 |
| yellow | 10 |
| other | 10 |

23. Rico created a table based on cars in the school parking lot. What is the probability that a randomly selected car in the school parking lot is white?

_____

24. What is the probability that a randomly selected car in the school parking lot is blue?

_____

25. Find the number of permutations of 6 different-colored light bulbs in a string of 6 sockets.

_____

26. Five different signal flags are available. How many different signals can be made by displaying all 5 on a vertical flagpole?

_____

27. How many different license plates can be made using 3 letters followed by 3 digits (0 through 9)? Letters and digits may be repeated.

_____

28. How many permutations of the letters A, B, C, D, E, and F have a vowel in the first and fifth positions?

_____

29. If all the letters are used, how many different permutations are possible using the letters of the word ELEMENT?

_____

30. If all the letters are used, how many different permutations are possible using the letters of the word LETTER?

_____

31. In how many ways can 5 people be seated around a circular table?

_____

32. In how many ways can 7 different charms be arranged on a charm bracelet?

_____

33. There are 20 people available to serve on a jury for a trial. In how many ways can a 12 person jury be selected from the people available?

_____

34. A summer basketball league with 11 teams decides to split into a division with 6 teams and a division with 5 teams. In how many ways can the league divide itself into two divisions?

_____

35. How many combinations can be formed from the letters in the word COUNT, taking them 2 at a time?

_____

36. If no letter is repeated, then how many 5-letter arrangements of the letters in the word LOGARITHMS consisting of 3 consonants and 2 vowels can be formed?

_____

37. How many committees of 2 Republicans and 3 Democrats can be chosen from a group of 7 Republicans and 10 Democrats?

_____

38. A water-control engineer must inspect a sample of 2 fuses from a package of 100. How many different samples can be chosen?

_____

39. What is the probability that neither of the first 2 rolls of a number cube will be a 1?

_____

40. Suppose you roll an honest number cube 4 times. What is the probability that at least once you will roll a 1? (Hint: first find the probability that you will never roll a 1)

_____

41. What would this formula be used to calculate?
$P(A) + P(B) - P(A \text{ and } B)$

_____

42. Two cards are drawn from a deck of 52 cards. What is the probability that either both are red or both are aces?

_____

43. $P(B \mid A) = $ _____

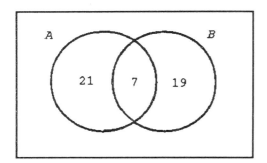

44. A family contains 2 children. Each child is equally likely to be a boy or a girl. Given that at least one child is a boy, what is the probability that the other child is also a boy? For the equation $P(E_2 \mid E_1)$, write out in words what $E_1$ and $E_2$ represent.

_____

_____

45. What is $P(E_1)$?

_____

46. What is $P(E_1 \text{ and } E_2)$?

_____

47. What is $P(E_2 \mid E_1)$?

_____

48. Jackie is standing on the corner tossing a nickel. She decides that she will toss the nickel 10 times, each time walking 1 block north if it comes up heads, and 1 block south if tails. Use a random number simulation to approximate the probability that she will end up 10 blocks north of her corner.

_____

49. Jackie is standing on the corner tossing a nickel. She decides that she will toss the nickel 10 times, each time walking 1 block north if it comes up heads, and 1 block south if tails. Use a random number simulation to approximate the probability that she will end up 6 blocks south of her corner.

_____

# Answers to Chapter Questions

1. Answer: c  Objective: 1A

2. Answer: d  Objective: 6a

3. Answer: c  Objective: 4a

4. Answer: b  Objective: 4a

5. Answer:
   a. $\frac{1}{2}$

   Objective: 1a

6. Answer:
   d. $\frac{1}{2}$

   Objective: 1a

7. Answer:
   b. $\frac{1}{2}$

   Objective: 1a

8. Answer: b. 720  Objective: 2a

9. Answer: d. 11,880  Objective: 2b

10. Answer: c. 36  Objective: 2b

11. Answer: b. 4200  Objective: 3a

12. Answer: a. 3360  Objective: 3a

13. Answer: b. 11!  Objective: 3b

14. Answer: a. 10  Objective: 4a

15. Answer: b. 10  Objective: 4a

16. Answer:
    b. $\frac{9}{25}$

    Objective: 5a

17. Answer:

    a. $\dfrac{5}{18}$

Objective: 5b


18. Answer:

    c. $\dfrac{P(E_1 \text{ and } E_2)}{P(E_1)}$

Objective: 6a


19. Answer: a. the conditional probability of event $A$ given event $B$.  Objective: 6a

20. Answer:

    $\dfrac{1}{13}$

Objective: 1a


21. Answer: experimental probabilities  Objective: 1b

22. Answer:

    $\dfrac{7}{10}$

Objective: 1b


23. Answer:

    $\dfrac{9}{40}$

Objective: 1b


24. Answer:

    $\dfrac{1}{40}$

Objective: 1b


25. Answer: 720  Objective: 2a

26. Answer: 120  Objective: 2a

27. Answer: 17,576,000  Objective: 2b

28. Answer: 48  Objective: 2b

29. Answer: 840  Objective: 3a

30. Answer: 180  Objective: 3a

31. Answer: 24  Objective: 3b

32. Answer: 720   Objective: 3b

33. Answer: $_{20}C_{12} = 125{,}970$   Objective: 4a

34. Answer: $_{11}C_6 = {_{11}C_5} = 462$   Objective: 4a

35. Answer: 10   Objective: 4a

36. Answer: $(_7P_3)(_3P_2) = 1260$   Objective: 4a

37. Answer: 2520   Objective: 4b

38. Answer: 4950   Objective: 4b

39. Answer:
$$\frac{25}{36}$$
Objective: 5a

40. Answer:
$$\frac{671}{1296} \approx 0.52$$
Objective: 5b

41. Answer: probability of $A$ or $B$ occurring   Objective: 5b

42. Answer:
$$\frac{55}{221} \approx 0.25$$
Objective: 5b

43. Answer:
$$\frac{1}{4}$$
Objective: 6a

44. Answer:
$E_1$ = event that at least 1 child is a boy
$E_2$ = event that the other child is a boy
Objective: 6a

45. Answer:
$$\frac{3}{4}$$
Objective: 6a

46. Answer:
$$\frac{1}{4}$$
Objective: 6a

47. Answer:

$$\frac{1}{3}$$

Objective: 6a

48. Answer:

$$\frac{1}{1024}$$

Objective: 7a

49. Answer:

$$\frac{45}{1024}$$

Objective: 7a

# CHAPTER 12

*QUANTITATIVE COMPARISON*

In the space provided, write:
a. if the quantity of Column A is greater than the quantity in Column B;
b. if the quantity in Column B is greater than the quantity in Column A;
c. if the two quantities are equal; or
d. if the relationship cannot be determined from the information given.

| Column A | Column B | Answer |
|----------|----------|--------|

1.

| the 1st term of an arithmetic sequence, if the 3rd term is 8 and the 16th term is 47 | the common difference of an arithmetic sequence, if the 3rd term is 8 and the 16th term is 47 | _____ |

2.

| the 5th term of the geometric sequence 2, -10, 50,... | the 6th term of the geometric sequence 2, -10, 50,... | _____ |

3. $$t_n = 11n - 8$$

| $t_{20}$ | $S_7$ | _____ |

4. $$t_n = .5(-4)^n$$

| $t_{12}$ | $S_{12}$ | _____ |

*MULTIPLE CHOICE   Circle the letter of the best answer choice.*

5. Identify the common difference of this arithmetic sequence.
   2, 7, 12, 17, ...
   a. 5          b. 7          c. 3          d. 8

6. Identify the common difference of this arithmetic sequence.
   $\frac{3}{2}, \frac{9}{4}, 3, \frac{15}{4}, \ldots$

   a. $\frac{5}{9}$          b. $\frac{4}{7}$          c. $\frac{3}{4}$          d. $\frac{5}{8}$

7. Identify the common ratio in this geometric sequence.
   4, -8, 16, -32, ...
   a. 2          b. -2          c. 4          d. -4

8. Identify the common ratio in this geometric sequence.
   $\frac{1}{x}, \frac{1}{x^2}, \frac{1}{x^3}, \ldots$

   a. $\frac{1}{x^5}$          b. $\frac{1}{x^4}$          c. $\frac{1}{x}$          d. $\frac{1}{x^2}$

9. Which of the following is a geometric sequence?
   a. 2, 4, 6, 8, ...      b. 10, 20, 30, 40, ...
   c. 27, 9, 3, 1, ...     d. 50, 100, 150, 200, ...

10. Find the sum of the first 20 terms of the series 5 + 8 + 11 + 14 + ...
    a. 670          b. 420          c. 510          d. 730

11. Find the 6th term of the geometric sequence 3, -15, 75, ...
    a. -375         b. 1875         c. -9375         d. none of these

12. Find the sum of the first 6 terms of the geometric series 3 + 15 + 75 + 375 + ...
    a. 15,780       b. 8234         c. 5720          d. 11,718

13. Find the sum of the series $\sum_{k=1}^{6} \left(\frac{1}{2}\right)^{k-1}$

    a. $\frac{75}{8}$      b. $\frac{63}{32}$      c. $\frac{21}{8}$      d. $\frac{31}{5}$

14. Find the sum of the infinite geometric series $1 + \frac{1}{3} + \frac{1}{9} + \frac{1}{27} + \cdots$

    a. $\frac{2}{3}$      b. $\frac{3}{2}$      c. $\frac{1}{6}$      d. $\frac{1}{18}$

15. Find the infinite sum of the geometric series if it exists given $t_1 = 4$ and $r = -\frac{1}{4}$.

    a. $\frac{13}{4}$      b. $\frac{12}{7}$      c. $\frac{16}{5}$      d. does not exist

16. Repeating decimals represent infinite geometric series. For example, 0.66666... represents 0.6 + 0.06 + 0.006 + ... Using 0.6 for $a_1$ and 0.1 for $r$, find the fractional notation for $0.\overline{21}$.

    a. $\frac{7}{33}$      b. $\frac{2}{21}$      c. $\frac{3}{43}$      d. $\frac{1}{17}$

17. Expand $(x^2 - 1)^5$.
    a. $-x^{10} - 5x^8 + 10x^6 - 10x^4 + 5x^2 - 1$
    b. $-x^{10} - 5x^8 + 10x^6 - 10x4 - 5x^2 - 1$
    c. $x^{10} - 5x^8 + 10x^6 - 10x^4 + 5x^2 - 1$
    d. $x^{10} + 5x^8 + 10x^6 - 10x^4 + 5x^2 - 1$

18. Find the 4th term of $(x - 3)^8$.
    a. $-1632x^5$      b. $-1512x^5$      c. $-1480x^4$      d. $-1323x^4$

19. Find the 12th term of $(a - 2)^{14}$.
    a. $-745,472a^3$   b. $-250,328a^4$   c. $-962,834a^5$   d. $-576,828a^3$

**SHORT ANSWER**   *Write the answer in the space provided.*

20. Identify the common difference of this arithmetic sequence.
    34, 27, 20, 13, 6, -1, -8, ...

    _____

21. Identify the common ratio in this geometric sequence.
    3, 6, 12, 24, ...

    _____

22. Identify the common ratio in this geometric sequence.
3, -6, 12, -24, ...

_____

23. Identify the common ratio in this geometric sequence.
$1, \frac{1}{2}, \frac{1}{4}, \frac{1}{8}, \cdots$

_____

24. Write a formula for the $n$th term of the arithmetic sequence 4, 7, 10, 13, ...

_____

25. Find the 301st term of the arithmetic sequence 4, 7, 10, 13, ...

_____

26. Find $a_7$ when $a_1 = 5$ and $d = 2$.

_____

27. The 3rd term of an arithmetic sequence is 8 and the 16th term is 47.
Construct the sequence.

_____

28. Find the sum of the first 14 terms of the arithmetic series $2 + 5 + 8 + 11 + 14 + 17 +$

_____

29. Find the sum of the series $\sum\limits_{k=1}^{13} (4x + 5)$.

_____

30. How many poles will be in a pile of telephone poles if there are 30 in the first layer, 29 in the second, and so on until there is one in the last layer?

_____

31. Find the 11th term of the geometric sequence 64, -32, 16, -8, ...

_____

32. A college student borrows $800 at 18% interest compounded annually. The loan is paid in full at the end of two years. How much has the student paid?

_____

33. Write a formula for the $n$th term of the geometric sequence 1, 3, 9,... Find $t_6$.

_____

34. Find the sum of the first 6 terms of the geometric series $3 + 6 + 12 + 24 + \ldots$

_____

35. Find the sum of the series $\sum\limits_{k=1}^{5} \left[\frac{1}{2}\right]^{k+1}$

36. A table tennis ball is dropped from a height of 16 ft and always rebounds $\frac{1}{4}$ of the distance of the previous fall. What distance does it rebound the 6th time?

37. Determine if the series has a sum: $1 - \frac{1}{2} + \frac{1}{4} - \frac{1}{8} + \frac{1}{16} + \cdots$

38. Determine if the series has a sum: $1 - 5 - 25 - 125 - \cdots$

39. Determine if the series has a sum: $1 + (-1) + 1 + (-1) + \cdots$

40. Find the infinite sum $\sum\limits_{k=1}^{\infty} (5(.5)^k)..$

41. Repeating decimals represent infinite geometric series. For example, 0.66666... represents $0.6 + 0.06 + 0.006 + \cdots$ Using 0.7 for $a_1$ and 0.1 for $r$, find the fractional notation for 0.7777.

42. How far up and down will a ball travel before stopping if it is dropped from a height of 12 m, and each rebound is $\frac{1}{3}$ of the previous distance?

43. The second last entry in a row of Pascal's Triangle is 17. Which row is it?

44. Write the 8th row of Pascal's Triangle.

45. Which row of Pascal's Triangle contains $_9C_4$?

46. Write the first 3 numbers in the 30th row of Pascal's Triangle.

47. Write the 12th number in the 17th row of Pascal's Triangle.

_____

48. If a family has 6 children, find the probability that the family has exactly 4 boys.

_____

49. If a fair coin is tossed 5 times, find the probability that heads occurs exactly once.

_____

50. Expand $(x^2 - 2y)^5$ .

_____

51. Expand $\left[2x + \dfrac{1}{y}\right]^4$ .

_____

52. Find the 7th term of $(4x - y^2)^9$ .

_____

53. Find the 6th term of $(y^2 + 2)^{10}$ .

_____

54. Find the middle term of the binomial expression $(2u - 3v^2)^{10}$ .

_____

# Answers to Chapter Questions

1. Answer: b  Objective: 2A

2. Answer: a  Objective: 3a

3. Answer: a  Objective: 2a

4. Answer: a  Objective: 3a

5. Answer: a. 5  Objective: 1a

6. Answer:
   c. $\frac{3}{4}$

   Objective: 1a

7. Answer: b. -2  Objective: 1b

8. Answer:
   c. $\frac{1}{x}$

   Objective: 1b

9. Answer: c. 27, 9, 3, 1, ...  Objective: 1b

10. Answer: a. 670  Objective: 2b

11. Answer: c. -9375  Objective: 3a

12. Answer: d. 11,718  Objective: 3b

13. Answer:
    b. $\frac{63}{32}$

    Objective: 3b

14. Answer:
    b. $\frac{3}{2}$

    Objective: 4a

15. Answer:
    c. $\frac{16}{5}$

    Objective: 4a

16. Answer:

a. $\dfrac{7}{33}$

Objective: 4a

17. Answer: c. $x^{10} - 5x^8 + 10x^6 - 10x^4 + 5x^2 - 1$      Objective: 6a

18. Answer: b. $-1512x^5$   Objective: 6b

19. Answer: a. $-745,472a^3$   Objective: 6b

20. Answer: -7   Objective: 1a

21. Answer: 2   Objective: 1b

22. Answer: -2   Objective: 1b

23. Answer:

$\dfrac{1}{2}$

Objective: 1b

24. Answer: $A_n = 4 + 3(n - 1)$   Objective: 2a

25. Answer: 904   Objective: 2a

26. Answer: 17   Objective: 2a

27. Answer: 2, 5, 8, 11, 14, ...   Objective: 2a

28. Answer: $S_{14} = 301$   Objective: 2b

29. Answer: $S_{13} = 429$   Objective: 2b

30. Answer: 465   Objective: 2b

31. Answer:

$\dfrac{1}{16}$

Objective: 3a

32. Answer: $1113.92   Objective: 3a

33. Answer: $t_n = 3^{n-1}$; $t_6 = 243$   Objective: 3a

34. Answer: 189   Objective: 3b

35. Answer:

$\dfrac{31}{64}$

Objective: 3b

36. Answer:

$$\frac{1}{256} \, ft$$

Objective: 3b

37. Answer:

$$r = -\frac{1}{2}, \ |r| < 1, \ \text{therefore the series has a sum}$$

Objective: 4a

38. Answer: $r = -5$, $|r| > 1$, therefore the series does not have a sum   Objective: 4a

39. Answer: $r = -1$, $|r| = 1$, therefore the series does not have a sum   Objective: 4a

40. Answer: 10   Objective: 4a

41. Answer:

$$\frac{7}{9}$$

Objective: 4a

42. Answer: 24 m   Objective: 4a

43. Answer: 17   Objective: 5a

44. Answer: 1      8      28      56      70      56      28      8      1   Objective: 5a

45. Answer: 9th   Objective: 5a

46. Answer: 1, 30, 435   Objective: 5a

47. Answer: 12376   Objective: 5a

48. Answer: 234375   Objective: 5a

49. Answer: 15625   Objective: 5a

50. Answer: $x^{10} - 10x^8 y + 40x^6 y^2 - 80x^4 y^3 + 80x^2 y^4 - 32y^5$   Objective: 6a

51. Answer:

$$16x^4 + 32\frac{x^3}{y} + 24\frac{x^2}{y^2} + 8\frac{x}{y^3} + \frac{1}{y^4}$$

Objective: 6a

52. Answer: $5376x^3 y^{12}$   Objective: 6b

53. Answer: $8064y^{10}$   Objective: 6b

54. Answer: $-1,959,552u^5 v^{10}$   Objective: 6b

# CHAPTER 13

## QUANTITATIVE COMPARISON

In the space provided, write:
a. if the quantity of Column A is greater than the quantity in Column B;
b. if the quantity in Column B is greater than the quantity in Column A;
c. if the two quantities are equal; or
d. if the relationship cannot be determined from the information given.

| Column A | Column B | Answer |
|---|---|---|

1.

| median starting salary for engineering | median starting salary for agriculture | _____ |
|---|---|---|

Engineering

Agriculture

20,000 25,000 30,000 35,000 40,000 45,000 50,000 55,000

Starting Salaries

2.

| standard deviation of the ages of students at your high school | standard deviation of the ages of faculty at your high school | _____ |
|---|---|---|

3.

| standard deviation of the weights of the slices of bread in a 1 pound loaf | mean for the weights of the slices of bread in a 1 pound loaf | _____ |
|---|---|---|

## MULTIPLE CHOICE  *Circle the letter of the best answer choice.*

4. Find the mean of the given test scores: {72, 80, 80, 82, 88, 90, 96}.
   a. 88          b. 84          c. 90          d. 72

5. Find the median of the given set of data: {61, 70, 84, 60, 65, 65, 73, 79, 73}.
   a. 65          b. 73          c. 70          d. 60

6. Find the mode of the given set of data: {85, 61, 68, 73, 91, 68, 93}.
   a. 68          b. 85          c. 73          d. 61

7. How many teens watch 2 to 3 hours of television each day?
   a. 2　　　　　b. 7　　　　　c. 4　　　　　d. 8

Television Viewing by Teens

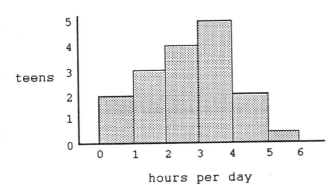

hours per day

8. Find the first quartile in the data set {0, 1, 2, 3, 4, 5, 6, 7, 8, 9}.
   a. 3　　　　　b. 2　　　　　c. 4　　　　　d. 5

9. Find the third quartile in the data set {0, 1, 2, 3, 4, 5, 6, 7, 8, 9}.
   a. 9　　　　　b. 1　　　　　c. 3　　　　　d. 7

10. The table shows the scores of a group of students on a 10-point
    multiple choice placement test. Use it to answer the next 4
    questions.

| Score | 3 | 4 | 5 | 6 | 7 | 8 | 9 | 10 |
|---|---|---|---|---|---|---|---|---|
| Number of Students | 5 | 3 | 5 | 2 | 7 | 6 | 1 | 1 |

    The total number of students taking the test is
    a. 52　　　　　b. 8　　　　　c. 30　　　　　d. 180

11. The range of the test scores is
    a. 7　　　　　b. 9　　　　　c. 8　　　　　d. 10

12. Whose scores were the outliers?
    a. The five students that scored a 3.
    b. The student that scored a 10.
    c. All those students whose scores were equal to the average.
    d. There were no outliers in this set of scores.

13. The equation for binomial probability distribution is
    a. $P(r) = {}_nC_r p^r q^{n-r}$
    b. $P(r) = {}_rC_n p^r q^{n-r}$
    c. $P(r) = {}_nC_r p^n q^r$
    d. $P(r) = {}_nC_r p^r q^{r-n}$

14. Find the percent of the total area under a normal curve between the mean and the given
    number of standard deviations from the mean at 2.50.
    a. 50.1%　　　　b. 36.3%　　　　c. 49.4%　　　　d. 61.4%

15. Suppose 100 geology students measure the mass of an ore sample. Due to human error and
    limitations in the reliability of the balance, not all the readings are equal. The results
    are found to closely approximate a normal distribution, with mean 37 g and standard
    deviation 1 g. Estimate the number of students reporting readings more than 35 g.
    a. 97 or 98　　　b. 77 or 78　　　c. 67 or 68　　　d. 57 or 58

16. The data below represents the number of written reports required of ten students. What are the mean and median number of written reports required?
    {6, 7, 8, 9, 10, 12, 15, 15, 20, 28}

    _____

17. If the data for a particular fish pond is {10, 15, 18, 19, 21, 24, 26, 27}, then what is the median?

    _____

18. Find the mode of the given set of data: {9, 8, 9, 8, 6, 7, 8, 9, 10, 8}.

    _____

19. The bar graph shows the frequencies of various scores received by students in an English 1005 course on a 10-point quiz. Construct a frequency table for the scores on the quiz.

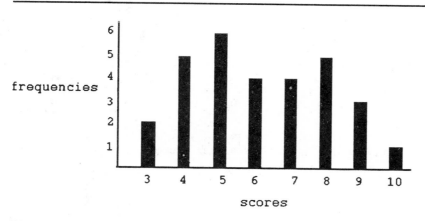

20. Find the mean, median and mode of the scores.

    _____

21. What percent of the students scored 8 or above?

    _____

22. Construct a histogram to show the data using the pulse rates of 60 people.

| Pulse Rates | Count |
|-------------|-------|
| 65-69       | 5     |
| 70-74       | 22    |
| 75-79       | 23    |
| 80-84       | 9     |
| 85-89       | 1     |

23. Using the histogram, list the data by the intervals.

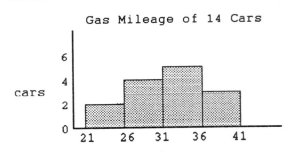

Gas Mileage of 14 Cars

cars

miles per gallon

24. Construct a histogram using the following data with the given intervals.

Diameters of Twenty Trees (cm)

| | | | | |
|---|---|---|---|---|
| 25 | 32 | 36 | 28 | 20 |
| 26 | 35 | 40 | 27 | 30 |
| 36 | 23 | 29 | 31 | 24 |
| 29 | 32 | 43 | 25 | 30 |

Intervals: 17-23; 24-30; 31-37; 38-44

25. Construct a histogram given the following set of data and intervals.
Miles Bicycled per day for Triathlon

Training

| | | | | |
|---|---|---|---|---|
| 25 | 30 | 22 | 32 | 37 |
| 34 | 15 | 25 | 39 | 24 |
| 32 | 18 | 30 | 35 | 32 |
| 20 | 25 | 31 | 28 | 30 |

Intervals: 15-19; 20-24; 25-29; 30-34; 35-39

26. The following data show the summer earnings of selected 17-year-olds with jobs. Make a stem-and-leaf table of the data. Use the first digit as the stem and the last 2 digits as the leaf.
Earnings: $280, $220, $275, $275, $290, $325, $375, $350, $325, $325, $455, $510

27. Daryl's class held a gift exchange. The teacher asked each student to record the price of the gift everyone brought. The data was placed in a stem-and-leaf table. Examine the table to answer the questions.

| Stem | Leaf |
|---|---|
| 2 | 25, 89, 75, 62, 99, 91, 95 |
| 3 | 50, 25, 15, 89, 95, 99, 97, 99, 50 |
| 4 | 45, 50, 75, 99, 99, 49, 15, 50 |
| 5 | 0, 0, 4 |

How many gifts cost more than $3.97?

28. How many students were in Daryl's class?

_____

29. Is $3.97 the mean, the median, or the mode for this set of data?

_____

30. Find the interquartile range in the data set {0, 1, 2, 3, 4, 5, 6, 7, 8, 9}.

_____

31. The heights in inches of 15 students in the Spanish Club are:  60, 62, 62, 65, 66, 66, 67, 67, 67, 68, 69, 70, 70, 72, 76.  Find $Q_1$, $Q_3$, and the interquartile range.

_____

32. Use the following data for the next 2 questions.

| Value | 9 | 10 | 11 | 12 | 13 | 14 | 15 | 16 | 17 | 18 | 19 | 20 |
|-------|---|----|----|----|----|----|----|----|----|----|----|----|
| Frequency | 3 | 5 | 7 | 4 | 12 | 10 | 13 | 11 | 15 | 13 | 11 | 4 |

| Value | 21 | 22 | 23 | 24 | 25 |
|-------|----|----|----|----|----|
| Frequency | 3 | 0 | 0 | 0 | 1 |

Find the five-number summary.

_____

33. Construct a box-and-whisker plot.

34. For the data set {1, 2, 3, 4, 5, 6, 7, 8, 9, 10}, find the standard deviation.

_____

35. Under what conditions would the standard deviation of a set of numbers be 0?

_____

36. Find the range, mean deviation, and the standard deviation of the set of scores {10, 11, 13, 14, 17}.

_____

37. Find the mean deviation and the standard deviation of the set {-3, -2, 1, 8}.

_____

38. A door-to-door salesperson makes a sale at 12% of the houses the person visits. Suppose the salesperson visits 20 houses in a day. Find the probability that the person makes exactly 4 sales in each day.

_____

39. Find the probability that the person makes no sales in a day.

_____

40. A disease has an 82% cure rate when treated with a certain drug. Suppose 16 people with the disease are given the drug. Find the probability that exactly 12 are cured.

_____

41. Find the probability that exactly 6 are not cured.

_____

42. A stockbroker gives correct advice on whether to buy or sell a stock 70% of the time. Suppose the stockbroker is asked about whether to buy or sell 30 stocks. Find the probability that the stockbroker gives the correct advice 20 times.

_____

43. Find the probability that the stockbroker gives the correct advice 25 times.

_____

44. About _____ % of the area under the normal distribution curve is within 2 standard deviations of the mean.

_____

45. Suppose 300 math students take a midterm exam and that their scores approximate a normal distribution. Find the number of scores falling within 1 standard deviation of the mean.

_____

46. Suppose 400 math students take a midterm exam and that their scores are normally distributed. Find the number of scores falling 2 standard deviations above the mean.

_____

47. Suppose 100 geology students measure the mass of an ore sample. Due to human error and limitations in the reliability of the balance, not all
the readings are equal. The results are found to closely approximate a normal distribution, with mean 37 g and standard deviation 1 g. Estimate the number of students reporting readings more than 37 g.

_____

48. Suppose 100 geology students measure the mass of an ore sample. Due to human error and limitations in the reliability of the balance, not all the readings are equal. The results are found to closely approximate a normal distribution, with mean 37 g and standard deviation 1 g. Estimate the number of students reporting readings below 36 g.

_____

49. Suppose 100 geology students measure the mass of an ore sample. Due to human error and limitations in the reliability of the balance, not all the readings are equal. The results are found to closely approximate a normal curve, with mean 37 g and standard deviation 1 g. Use the symmetry of the normal curve and the empirical rule to estimate the number of students reporting readings within 1 g of the mean.

_____

176

# Answers to Chapter Questions

1. Answer: a  Objective: 3B

2. Answer: b  Objective: 4b

3. Answer: b  Objective: 4b

4. Answer: b. 84  Objective: 1a

5. Answer: c. 70  Objective: 1a

6. Answer: a. 68  Objective: 1a

7. Answer: c. 4  Objective: 2a

8. Answer: b. 2  Objective: 3a

9. Answer: d. 7  Objective: 3a

10. Answer: c. 30  Objective: 4a

11. Answer: a. 7  Objective: 4a

12. Answer: d. There were no outliers in this set of scores.  Objective: 4a

13. Answer: a. $P(r) = {}_nC_r p^r q^{n-r}$  Objective: 5a

14. Answer: c. 49.4%  Objective: 6a

15. Answer: a. 97 or 98  Objective: 6a

16. Answer: 13; 11  Objective: 1a

17. Answer: 20  Objective: 1a

18. Answer: 8  Objective: 1a

19. Answer:

| Score | Frequency |
|-------|-----------|
| 3     | 2         |
| 4     | 5         |
| 5     | 6         |
| 6     | 4         |
| 7     | 4         |
| 8     | 5         |
| 9     | 3         |
| 10    | 1         |

Objective: 1b

20. Answer: 6.17; 6.5  Objective: 1b

21. Answer: 30%   Objective: 1b

22. Answer:

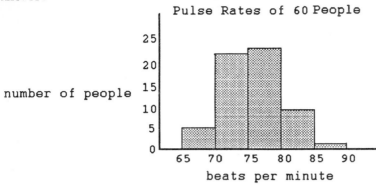

Pulse Rates of 60 People

number of people

beats per minute

Objective: 2a

23. Answer:

| Gas Mileage | Count |
| --- | --- |
| 21-25 | 2 |
| 26-30 | 4 |
| 31-35 | 5 |
| 36-40 | 3 |

Objective: 2a

24. Answer:

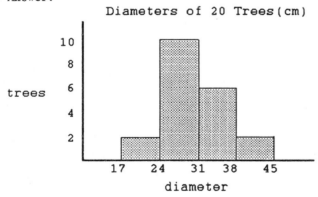

Diameters of 20 Trees(cm)

trees

diameter

Objective: 2a

25. Answer:

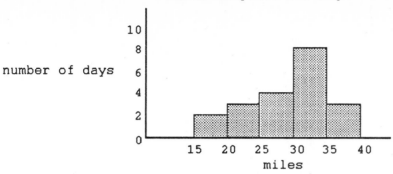

Objective: 2a

26. Answer:

| Stem | Leaf |
|------|------|
| 2 | 80, 20, 75, 75, 90 |
| 3 | 25, 75, 50, 25, 25 |
| 4 | 55 |
| 5 | 10 |

Objective: 2b

27. Answer: 13   Objective: 2b

28. Answer: 27   Objective: 2b

29. Answer: the median   Objective: 2b

30. Answer: 5   Objective: 3a

31. Answer: $Q_1 = 65; Q_3 = 70;$ $IQR = 5$   Objective: 3a

32. Answer: min = 9, $Q_1 = 13$, median = 16, $Q_3 = 18$, max = 25   Objective: 3b

33. Answer:

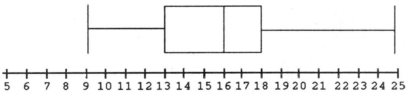

Objective: 3b

34. Answer: about 2.87   Objective: 4b

35. Answer:
When the deviations are all 0 or the numbers in the data set are all equal (and therefore equal to their average).

Objective: 4b

36. Answer: 7; 2; 2.45   Objective: 4b

37. Answer:
$\frac{7}{2}$, 4.30

Objective: 4b

38. Answer:   13   Objective: 5a

39. Answer:   .08   Objective: 5a

40. Answer:   .18   Objective: 5a

41. Answer:   .04   Objective: 5a

42. Answer:   .14   Objective: 5a

43. Answer:   .05   Objective: 5a

44. Answer: 95   Objective: 6a

45. Answer: 204   Objective: 6a

46. Answer: 10   Objective: 6a

47. Answer: 50   Objective: 6a

48. Answer: 16   Objective: 6a

49. Answer: 68   Objective: 6a

# CHAPTER 14

## QUANTITATIVE COMPARISON

In the space provided, write:
a. if the quantity of Column A is greater than the quantity in Column B;
b. if the quantity in Column B is greater than the quantity in Column A;
c. if the two quantities are equal; or
d. if the relationship cannot be determined from the information given.

| Column A | Column B | Answer |
|---|---|---|

1.

| $\sin^2 x + \cos^2 x$ | $\sec^2 x - \tan^2 x$ | _____ |
|---|---|---|

2.

$$\cot x = \frac{8}{15} \text{ and } x \text{ is the third quadrant}$$

| $\tan x$ | $\sec x$ | _____ |
|---|---|---|

3.

| $\sin\left[\dfrac{\pi}{3} + \dfrac{\pi}{4}\right]$ | $\cos\left[\dfrac{\pi}{4} - \dfrac{\pi}{6}\right]$ | _____ |
|---|---|---|

**MULTIPLE CHOICE** *Circle the letter of the best answer choice.*

4. If $c = 916$, $m\angle A = 15° \ 40'$ and $m\angle B = 60° \ 30'$, then find $a$.
   a. 255        b. 370        c. 162        d. 574

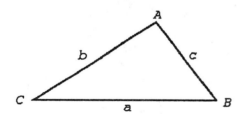

5. If $m\angle A = 35°$, $m\angle C = 115°$, and $b = 250$, then what length is $c$?
   a. 507        b. 453        c. 232        d. 847

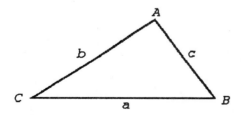

6. A ship at sea is sighted at two lighthouses, $A$ and $B$, on shore. The
   lighthouses are 15 mi apart. The measure of the angle at $A$ between $\overline{AB}$ and the ship is 41°
   40'. The angle at $B$ is 36° 10'. Find the distance to the nearest tenth of a mile from $A$ to
   the ship.
   a. 14.2 mi        b. 8.3 mi        c. 10.3 mi        d. 9.1 mi

7. A pilot approaching a 10,000-ft runway finds that the angles of depression of the ends of the runway are 12° and 15°. How far is the pilot from the nearest end of the runway?
   a. 27,800 ft     b. 39,700 ft       c. 42,300 ft       d. 44,200 ft

8. On one bank of a river are vertices $A$ and $B$ and on the opposite bank is vertex $C$. The length of $\overline{AB}$ is 200 ft and $m\angle A = 33°$ and $m\angle B = 63°$. Find the distance between points $C$ and $A$.
   a. 202 ft       b. 310 ft         c. 252 ft         d. 179 ft

9. If $a = 10$, $b = 9\sqrt{3}$, and $m\angle C = 86°$, then find $c$.
   a. 17.96       b. 16.27        c. 13.2          d. 14.81

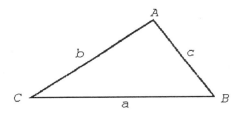

10. Find $m\angle B$ if $b = 9$, $c = 15$, and $m\angle A = 94$.
    a. 54°         b. 27°         c. 42°          d. 30°

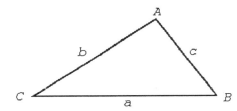

11. Find $m\angle C$ if $a = b = 42.5$ and $c = 83.1$.
    a. 123°       b. 72°         c. 98°          d. 156°

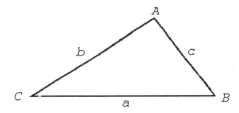

12. If $a = 1.6$, $b = 2.5$, and $c = 1.9$, find the smallest angle.
    a. 39.8°      b. 42.6°       c. 27.2°      d. 29.8°

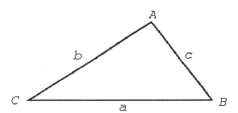

13. Find $\cos \theta$ and $\sin \theta$.

a. $\dfrac{2}{3}$, $-\dfrac{3}{5}$     b. $\dfrac{6}{7}$, $-\dfrac{3}{5}$     c. $\dfrac{2}{3}$, $-\dfrac{3}{5}$     d. $\dfrac{4}{5}$, $-\dfrac{3}{5}$

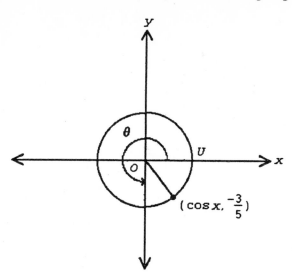

14. Given that $\csc x = \dfrac{2}{\sqrt{3}}$ and $x$ is in the second quadrant, find $\cos x$.

a. 0.5     b. $\dfrac{\sqrt{3}}{2}$     c. $2\sqrt{3}$     d. 1

15. Which of the following is true?

a. $\cot x = \dfrac{\cos x}{\sec x}$     b. $\sec x = \dfrac{1}{\sin x}$

c. $1 + \tan^2 x = \left[\dfrac{1}{\cos^2 x}\right]$     d. $\sin^2 x + 1 = \cos^2 x$

16. Given that $\sin x = -\dfrac{5}{13}$, and $x$ is a fourth-quadrant number, find $\cos x$.

a. $\dfrac{12}{13}$     b. $\dfrac{5}{6}$     c. $-\dfrac{3}{7}$     d. $\dfrac{14}{17}$

17. Given that $\cos x = \dfrac{-\sqrt{5}}{3}$, and $x$ is a second-quadrant number, find $\sin x$.

a. $\dfrac{5}{6}$     b. $-\dfrac{3}{7}$     c. $\dfrac{2}{3}$     d. $\dfrac{6}{11}$

18. Find the exact value of the expression $\cos\left[-\dfrac{2\pi}{3}\right]$.

a. $-\dfrac{\sqrt{2}}{2}$     b. $-\dfrac{1}{2}$     c. $\dfrac{2}{3}$     d. $\dfrac{\sqrt{2}}{2}$

19. Find the exact value of the expression $\sin\left[\dfrac{2\pi}{3} - \pi\right]$.

a. $-\dfrac{1}{2}$     b. $\dfrac{1}{2}$     c. $-1$     d. $-\dfrac{\sqrt{3}}{2}$

20. Find the exact solutions for $4 \cos^2 x = 1$ for $0 \leq x < 2\pi$.

    a. $\dfrac{\pi}{3}, \dfrac{2\pi}{3}, \dfrac{4\pi}{3}, \dfrac{5\pi}{3}$              b. $-\dfrac{1}{2}, 1$

    c. $\dfrac{\pi}{2}, \dfrac{2\pi}{3}, \dfrac{3\pi}{4}$              c. $0, -1, 1$

***SHORT ANSWER***    *Write the answer in the space provided.*

21. Find $a$ if $m\angle A = 38°$, $m\angle C = 54°$, and $b = 20.4$.

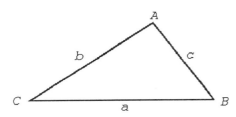

22. Find $c$ if $m\angle A = 100°$, $m\angle C = 38.5°$, and $a = 0.120$.

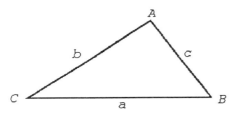

23. If $a = 1.50$, $m\angle B = 32° \, 30'$ and $b = 2.3$, then find $m\angle A$.

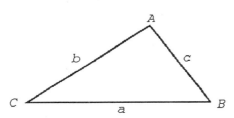

24. Find $m\angle A$ if $m\angle B = 40°$, $b = 16$, and $a = 24$.

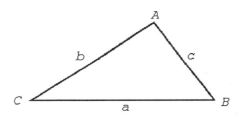

25. While on opposite ends of a beach 7.4 mi in length which runs due south and north, two people see a ship on fire. Their lines of sight to the ship have bearings S 59° E and N 40° E. How far is the ship from the nearest person?

26. Two weather-radar stations are located 18 mi apart from each other. They spot the same thunderstorm. If each locates the storm at an angle of 110° and 52° respectively, from the other station, how far is the storm from the station nearest it?

_____

27. An isosceles triangle has a base length of 20 in. If the vertex angle of the triangle measures 30°, find the perimeter of the triangle.

_____

28. If $b = 7$, $c = 10$, and m∠$A = 30°$, then find $a$.

_____

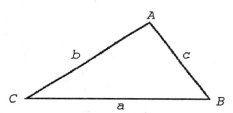

29. A tunnel is to be dug from point $A$ to point $B$. The distances from a third point, $C$, to $A$ and $B$ are 2.63 mi and 1.84 mi respectively, and m∠$ACB = 52.2°$. How long will the tunnel be?

_____

30. Find m∠$C$, given that $b = 7$, $c = 10$, and m∠$A = 30^0$.

_____

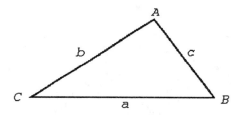

31. Find m∠$B$, given that $a = 4$, $b = 5$, and m∠$C = 126°$.

_____

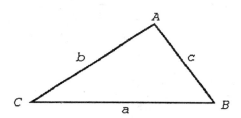

32. If $a = 12$, $b = 15$, and $c = 20$, find $m\angle B$.

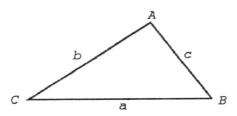

33. Find the largest angle, if $a = 12.6$, $b = 15.5$, and $c = 25.0$.

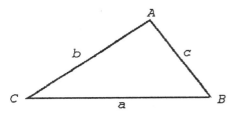

34. If $a = 24$, $b = 12$, and $c = 15$, what is $m\angle A$?

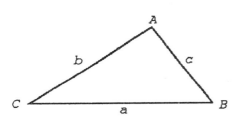

35. A triangular lot has frontages of 30 m and 50 m on two streets, and its third side is 60 m long. At what angle do the streets intersect?

36. Find $\sec\theta$ and $\csc\theta$.

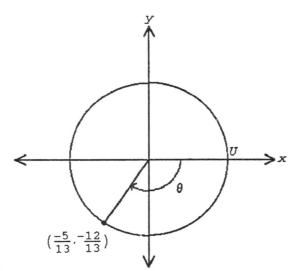

$\left(\dfrac{-5}{13}, \dfrac{-12}{13}\right)$

37.

Given that $\tan x = \sqrt{3}$ and $x$ is in the first quadrant, find $\sec x$.

38.

Find the exact value of the expression $\sin\left[-\dfrac{5\pi}{4}\right]$.

39.

Find the exact value of the expression $\cos\left[\dfrac{\pi}{2} + \dfrac{\pi}{4}\right]$.

40. Show the equation is true.

$\sin\dfrac{\pi}{3} = 2\sin\dfrac{\pi}{6}\cos\dfrac{\pi}{6}$

41. Show the equation is true.

$\cos\dfrac{4\pi}{3} = \cos^2\dfrac{2\pi}{3} - \sin^2\dfrac{2\pi}{3}$

42. Show the equation is true.

$\cos\dfrac{5\pi}{3} = 2\cos^2\dfrac{5\pi}{6} - 1$

43. Show the equation is true.

$\cos\dfrac{3\pi}{2} = 1 - 2\sin^2\dfrac{3\pi}{4}$

44. Find the exact value for $\sin 150°$ using trigonometric identities.

45. Find the solutions for $\sin x + \cos x = 0$ for $0 \le x < 2\pi$.

46. Find the exact solutions for $\sin x = \cos x$ for $0 \le x < 2\pi$.

47. Find the solutions for $\sin 2x + \cos 2x = 0$ for $0 \le x < 2\pi$.

48. Find the solutions for $\sin 2x + \cos x = 0$ for $0 \le x < 2\pi$.

49. Find the solutions for $\cos 2x = 3 \cos x + 1$ for $0° \leq x < 360°$ or $0 \leq x < 2\pi$.

_____

50. Find the solutions for $3 \sin x = 1 + \cos 2x$ for $0 \leq x < 2\pi$.

_____

# *Answers to Chapter Questions*

1. Answer: c  Objective: 3A

2. Answer: a  Objective: 3a

3. Answer: c  Objective: 4a

4. Answer: a. 255  Objective: 1a

5. Answer: b. 453  Objective: 1a

6. Answer: d. 9.1 mi  Objective: 1a

7. Answer: b. 39,700 ft  Objective: 1a

8. Answer: d. 179 ft  Objective: 1a

9. Answer: a. 17.96  Objective: 2a

10. Answer: d. 30°  Objective: 2a

11. Answer: d. 156°  Objective: 2b

12. Answer: a. 39.8°  Objective: 2b

13. Answer:
    d. $\frac{4}{5}$, $-\frac{3}{5}$

    Objective: 3a

14. Answer: a. 0.5  Objective: 3a

15. Answer:
    c. $1 + \tan^2 x = \left[\dfrac{1}{\cos^2 x}\right]$

    Objective: 3a

16. Answer:
    a. $\dfrac{12}{13}$

    Objective: 3a

17. Answer:
    c. $\dfrac{2}{3}$

    Objective: 3a

18. Answer:
    b. $-\dfrac{1}{2}$

    Objective: 4a

19. Answer:
    d. $-\dfrac{\sqrt{3}}{2}$

    Objective: 4a

20. Answer:
    a. $\dfrac{\pi}{3}, \dfrac{2\pi}{3}, \dfrac{4\pi}{3}, \dfrac{5\pi}{3}$

    Objective: 5a

21. Answer: $a = 12.57$   Objective: 1a

22. Answer: $c = 0.0759$   Objective: 1a

23. Answer: $m\angle A = 20.51° = 20° 31'$   Objective: 1a

24. Answer: 74.6° or 105.4°   Objective: 1a

25. Answer: about 4.8 mi   Objective: 1a

26. Answer: 46 mi   Objective: 1a

27. Answer: 97.3 in.   Objective: 1a

28. Answer: $a = 5.268$   Objective: 2a

29. Answer: 2.09 mi   Objective: 2a

30. Answer: $m\angle C = 71.63°$   Objective: 2a

31. Answer: $m\angle B = 30.24°$   Objective: 2a

32. Answer: $m\angle B = 48.35°$   Objective: 2b

33. Answer: 125.4°   Objective: 2b

34. Answer: $m\angle A = 125.1°$   Objective: 2b

35. Answer: 93.8°   Objective: 2b

36. Answer:
    $-\dfrac{13}{5}, -\dfrac{13}{12}$

    Objective: 3a

37. Answer: 2   Objective: 3a

190

38. Answer:

$$\frac{\sqrt{2}}{2}$$

Objective: 4a

39. Answer:

$$-\frac{\sqrt{2}}{2}$$

Objective: 4a

40. Answer:

$$2 \sin \frac{\pi}{6} \cos \frac{\pi}{6} = 2 \cdot \frac{1}{2} \cdot \frac{\sqrt{3}}{2} = \frac{\sqrt{3}}{2} = \sin \frac{\pi}{3}$$

Objective: 4a

41. Answer:

$$\cos^2 \frac{2\pi}{3} - \sin^2 \frac{2\pi}{3} = \left[-\frac{1}{2}\right]^2 - \left[\frac{\sqrt{3}}{2}\right]^2 = \frac{1}{4} - \frac{3}{4} = -\frac{1}{2} = \cos \frac{4\pi}{3}$$

Objective: 4a

42. Answer:

$$2 \cos^2 \frac{5\pi}{6} - 1 = 2 \left[-\frac{\sqrt{3}}{2}\right]^2 - 1 = \frac{3}{2} - 1 = \frac{1}{2} = \cos \frac{5\pi}{3}$$

Objective: 4a

43. Answer:

$$1 - 2 \sin^2 \frac{3\pi}{4} = 1 - 2\left[\frac{\sqrt{2}}{2}\right]^2 = 1 - 1 = 0 = \cos \frac{3\pi}{2}$$

Objective: 4a

44. Answer:

$$\frac{1}{2}$$

Objective: 4a

45. Answer:

$$\frac{3\pi}{4}, \frac{7\pi}{4} \approx 2.356, \ 5.498$$

Objective: 5a

191

46. Answer:
$\frac{\pi}{4}, \frac{5\pi}{4}$ or 0.785, 3.927

Objective: 5a

47. Answer:
$\frac{3\pi}{8}, \frac{7\pi}{8}, \frac{11\pi}{8},$ or $\frac{15\pi}{8} \approx$ 1.178, 2.749, 4.320, 5.890

Objective: 5a

48. Answer:
$\frac{\pi}{2}, \frac{7\pi}{6}, \frac{3\pi}{2},$ or $\frac{11\pi}{6}$

Objective: 5a

49. Answer:
$\frac{2\pi}{3}$ or $\frac{4\pi}{3} \approx$ 2.094 or 4.189

Objective: 5a

50. Answer:
$\frac{\pi}{6}, \frac{5\pi}{6} \approx$ 0.524, 2.618

Objective: 5a